The Private Life of
SPIDERS

The Private Life of
SPIDERS

Paul Hillyard

NEW HOLLAND

Published in 2007 by New Holland Publishers (UK) Ltd
London • Cape Town • Sydney • Auckland

www.newhollandpublishers.com

Garfield House, 86–88 Edgware Road, London W2 2EA,
United Kingdom

80 McKenzie Street, Cape Town 8001, South Africa

Unit 1, 66 Gibbes Street, Chatswood, NSW 2067, Australia

218 Lake Road, Northcote, Auckland, New Zealand

10 9 8 7 6 5 4 3 2 1

ISBN 978 1 84537 690 1

Editorial Director: Jo Hemmings
Senior Editor: James Parry
Assistant Editor: Giselle Osborne
Design: Alan Marshall
Production: Hazel Kirkman

Origination by Pica Digital PTE Ltd, Singapore
Printed and bound in Singapore by Tien Wah Press

Photographs: front cover: fishing spider on cornflower; back
cover: Pink-toed Tarantula seen through leaf; front flap: orb web
spider; back flap: Regal Jumping Spider; page 1: male black
widow spider; page 2: crab spider on Bluebell; page 3:
completed orb web; pages 4/5: male green lynx spider signalling
to female in courtship display; page 6: green lynx spider on
Pitcher plant; page 7 (from top to bottom): house spider
(Tegenaria domestica) on wall; Cucumber Green Spider on Iris;
jumping spider nesting in silk pouch.

CONTENTS

CHAPTER 1
BACKGROUND TO SPIDERS

To delve into the private life of spiders you'll have to keep your wits about you; living alongside us, these spectacular creatures inhabit places that we may otherwise overlook. Keep your eyes open and spiders that you would never normally see emerge from their hiding places. Some are on the move hunting their prey, while others are busy building their webs. To detect as many species as possible, you'll need to look in all directions; up to the foliage, straight ahead on the tree trunks, and down on the ground. You may stumble across two male spiders competing to court a female in her web, while at the same time an unrelated species has invaded the web to steal her food. It's fascinating to observe what goes on in this other world.

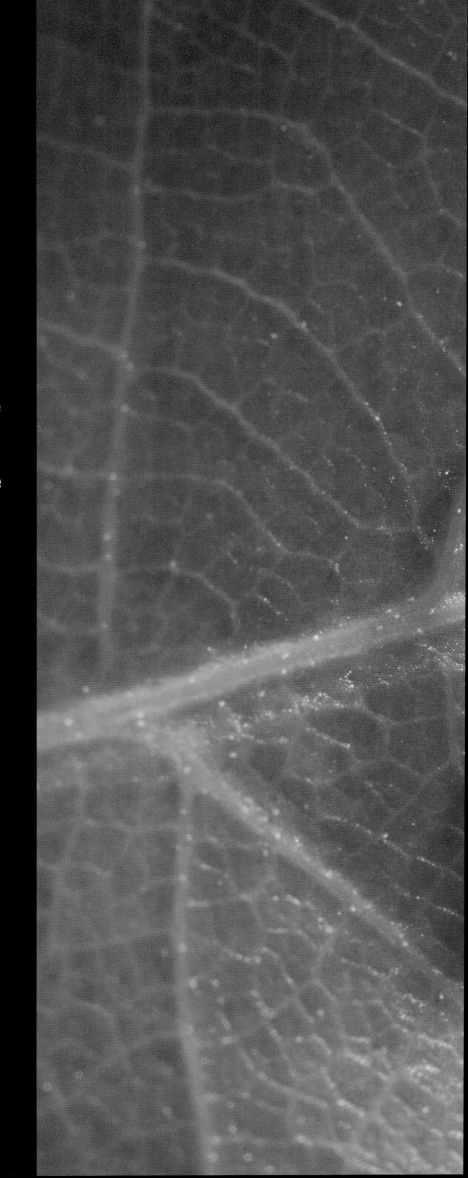

SPIDERS HAVE SPENT OVER 300 MILLION YEARS SHARPENING THEIR SKILLS FOR STAYING ALIVE AND PRODUCING NEW GENERATIONS OF SPIDERLINGS. THEY APPEARED ABOUT 100 MILLION YEARS BEFORE THE BACK-BONED ANIMALS AND, WITH THEIR FAST GENERATION TIMES, THEY HAVE EVOLVED MUCH MORE QUICKLY. THEY HAVE HAD PLENTY OF TIME TO DEVELOP COMPLEX FORMS OF BEHAVIOUR.

Previous: *This spider (Clubiona sp.) does not make a web, instead it hunts by stealth at night.*

Below: *Ladybird spider (Eresus, male). One of the prettiest of all spiders.*

Like all living things, spiders have to face life's problems, such as finding food and mates, producing offspring, and fending off danger. It is the many solutions that spiders have developed to counter these problems, and their highly complex skills, that make them so interesting. Take for instance their diverse strategies for catching prey. As well as devising all manner of ingenious traps in the form of webs, they also employ a host of other methods of attack. These include stalking, chasing, jumping, ambushing, fishing, robbing, spitting, masquerading as other animals, and even luring prey by mimicking their chemical scents (pheromones).

In common with insects and crustaceans, spiders are classed as arthropods. All have jointed legs and a more or less hard external armour, or exoskeleton. Together with their eight-legged relatives, the scorpions, harvestmen, mites, ticks and others, spiders form the class Arachnida. Spiders alone represent the order Araneae and are classified in a family tree that numbers three suborders and 110 families. The three suborders are: the 'true' spiders (Araneomorphae), which represent more than 90 per cent of all species; the tarantulas and typical trapdoor spiders (Mygalomorphae); and the giant trapdoor ('living fossil') spiders (Liphistiomorphae). Today, scientists have identified 38,000 different species of spiders worldwide. It is possible that a similar number remains still to be discovered.

Some other arachnids also produce silk, but spiders have exploited it much more, which helps explain their relative dominance. In addition, attributes such as jaws which inject venom, legs with hydraulic extension permitting fast movements and jumps, direct sperm transfer from male to female, great powers of dispersal and, in some cases, highly developed eyesight, have all contributed to the success of spiders as a group.

The largest species, South American tarantulas or bird-eating spiders, reach a body length of 10cm and may have a leg span of up to 27cm. But the smallest species are really tiny – less than one millimetre body length when adult. Nevertheless, all are predators. And all, with the exception of one family, use venom to paralyse their prey. Biting in self defence appears to be only a secondary purpose.

ECOLOGY AND DISTRIBUTION

Spiders are found virtually everywhere: in the house, in the garden, on lamp-posts, in forests, in caves, and in most other terrestrial habitats throughout the world. A few species live on or under water and some even in the marine tidal zone. While the greatest diversity of species occurs in tropical rainforest, spiders are also very well represented in temperate woodland, heathland and grassland. They occupy many different niches. For example, if a woodland is divided into its various zones (ground layer, field layer, shrub layer, woody zone and canopy) we find that a different community of spiders occurs in each zone. Spiders thrive wherever there is rich vegetation and plenty of insects or other arthropods. For example, they will move into bushes that are in flower and attracting plenty of insects. However, many

"THE LARGEST SPECIES, SOUTH AMERICAN TARANTULAS OR BIRD-EATING SPIDERS, REACH A BODY LENGTH OF 10CM AND MAY HAVE A LEG SPAN OF UP TO 27CM"

species are specialists able to survive in barren and hostile environments such as deserts and mountain tops. Perhaps the only part of the world that spiders have not managed to colonize is the ice cap of Antarctica. One of the most remarkable things about spiders is that some live in the most unpromising habitats, such as underground cable ducting – places that seem to be distinctly lacking in insect prey.

Below: Orb-web spider wrapping prey (another spider).

In favourable situations, such as unspoilt meadows and woodland glades, spiders can be present in great numbers. The arachnologist W. S. Bristowe calculated in the 1930s that, at certain seasons, a meadow in south-east England may contain more than two million spiders to the acre. He estimated, for the country as a whole, that the weight of insects consumed annually by spiders exceeded the weight of its human inhabitants. According to Bristowe this was a conservative estimate but now, decades later and due to the loss of meadows and the general use of insecticides, not to mention the increased weight of humans, it is probably no longer true.

FOSSIL SPIDERS AND THE EARLIEST WEBS

Spiders are an ancient group that first appeared during the Devonian period, almost 400 million years ago. By the Carboniferous period (300 mya), when insects were still relatively little developed, many highly evolved spiders already existed. Some 20 fossil species have been described from this period, most of them remarkably like the surviving liphistiomorph trapdoor spiders of Southeast Asia. In the Mesozoic era (240 to 65 mya), spider fossils are extremely rare but the first specimens, of what are believed to have been araneomorphs, have been found from the Triassic period (225 mya). In addition, a fossil money spider (family Linyphiidae) has been found in Cretaceous amber (fossil tree resin) from the Middle East (125 mya). Since then, increasing numbers of well-preserved spiders similar to those alive today have been found, particularly in amber from Southeast Asia (90 mya), the Baltic (40 mya), the Caribbean (20 mya) and elsewhere.

Though there are no fossil webs, the earliest evidence for the production of silk comes from a fossil spinneret found in Devonian shale (about 380 million years old) from New York State. It is claimed to be the oldest spider fossil. The spinneret is a single segment carrying 20 spigots. It resembles the spinnerets of the giant trapdoor spiders of Southeast Asia. This spider uses silk to line the burrow and make a door with trip lines extending out.

Opposite: Crab spider (Misumena) *with bee prey.*

Left: *Spider fossilized in amber, approximately 40 million years old.*

DORSAL VIEW OF SPIDER

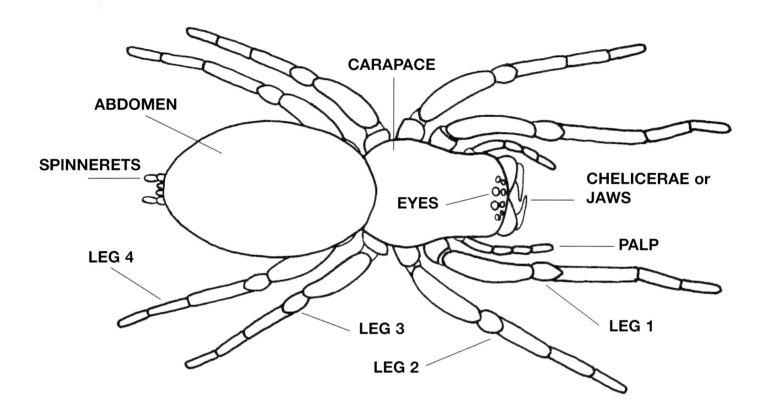

CARAPACE

ABDOMEN

SPINNERETS

CHELICERAE or JAWS

EYES

PALP

LEG 4

LEG 3

LEG 1

LEG 2

VENTRAL VIEW OF SPIDER

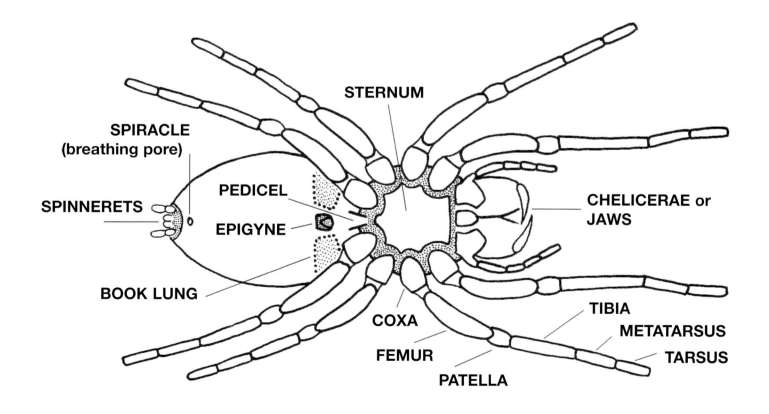

STERNUM

SPIRACLE (breathing pore)

PEDICEL

SPINNERETS

EPIGYNE

CHELICERAE or JAWS

BOOK LUNG

COXA

TIBIA

METATARSUS

FEMUR

TARSUS

PATELLA

SPIDER ANATOMY

The ancestry and anatomy of spiders are different from those of insects. Spiders have eight legs not six, and their bodies are divided into two parts rather than three. They have no antennae. The front part of a spider, covered by a carapace, consists of a united head and thorax called a cephalothorax. It contains the brain and stomach and carries the eight legs, a pair of jaws, a pair of palps (leg-like feelers on either side of the jaws), and the eyes – usually eight in number but sometimes reduced to six, or fewer. The way the eyes are arranged and developed is an important characteristic of each family. Spiders' eyes are generally classed as 'simple ocelli'; they are different from the 'compound' eyes found in many insects. In the case of the adult male spider, the palps carry a pair of 'accessory' sex organs used for the transfer of sperm.

The rear section of a spider is the abdomen. It contains the heart, digestive tract, silk glands, reproductive organs, and respiratory system. The abdomen is usually soft and round or elongate. In some species, it is strangely shaped and ornamented, and may even be hard and spiny. Some spiders are brightly coloured but most have sombre patterns to provide camouflage, particularly in the case of females. At the end of abdomen lie the spinning organs (spinnerets) (see diagram on page 14). These produce fine lines of silk which emerge through microscopic nozzles (spigots). The spinnerets, numbering two, four or six, are each shaped like a teat or finger. Both the male and female have their genital openings on the underside of the abdomen. The front and rear halves of the body are joined by the thinnest of waists (pedicel), through which pass the nerve cord, aorta and digestive tract.

The chelicerae, or jaws, with their movable fangs connected to venom glands, are the spider's offensive weapons. They are used not only for attack and defence, but also for manipulation in many tasks. Nursery-web spiders use them to carry egg cocoons; and trapdoor spiders use them, furnished with extra teeth, to dig burrows. The way the jaws move is an important characteristic in spider classification. In the suborder Mygalomorphae, for example in the tarantulas, the chelicerae strike downwards like two parallel pickaxes. In the more highly evolved suborder Araneomorphae, the jaws close together horizontally, like a pair of pincers.

In most cases the visual acuity of spiders is poor. They mostly 'listen' to the world around them through vibrations on the

Left: Jumping spider (Thiodina sylvana) showing eyes, jaws and palps. Jumping spiders have exceptional eyesight.

ground, in the air, via their webs, or across the surface of water. Their legs have many sensitive hairs and tiny slits, connected to nerves, which detect the slightest vibrations. Each leg ends in claws and may have, particularly in hunting spiders, a dense brush of hairs which allows adhesion on vertical and overhanging surfaces such as the underside of leaves. Magnified, these hairs can be seen to split into many thousands of fine extensions (*end feet*), providing a huge number of contact points to increase the forces of adhesion. Spiders are meticulous in cleaning legs, palps and chelicerae. They are far from being dirty creatures.

Spiders breathe air via two separate systems: book lungs and tracheae. A particular spider will have one or both of these systems. Araneomorphs usually have one pair of book lungs, whereas the mygalomorphs and liphistiomorphs have two pairs. A book lung contains a number of overlapping folds or leaves, which bring blood into contact with air. The tracheae, like those of insects, are a system of fine, branching tubes which carry oxygen throughout the body. In some small spiders with high metabolic rates and a tendency to desiccate, all book lungs have been replaced by the more efficient tracheae. However, it may be an advantage to have both of the independent systems, as in the case of wolf spiders, which are among the most active.

LIFE CYCLE

Depending on the climate, eggs laid in the autumn or dry season usually hatch in the following spring or wet season. Most spiders have a life cycle that lasts one year. Some, however, live for two to five years and some female tarantulas and trapdoor spiders live as many as 25 years (at least in captivity). At the other end of the scale, fast-living, tropical jumping spiders may live no more than a few months and pass through several generations in a year. Newly hatched spiders resemble small versions of the fully grown adults. In its growth from spiderling to adult, a spider undergoes approximately four to twelve moults, each time shedding the old rigid skin, or exoskeleton, for a new, larger one. Males, being smaller, usually have fewer moults than females and they tend to mature earlier.

Right: *Spider moulting skin. It is in the final process of extracting its legs from the old exoskeleton.*

Egg-laying and maternal care

The hazards of life are so great that most females produce tens, hundreds or even thousands of eggs to ensure survival of the species. Cave spiders of the genus *Telema* lay only a single egg, but tropical wandering spiders, such as the American *Cupiennius salei*, lay as many as 2,500. The abdomen of a female can swell considerably to accommodate large numbers of eggs. But after laying a big batch, the energy reserves are depleted and the body is noticeably shrunken. For protection against adverse conditions and egg parasites, the eggs are usually wrapped in a silk cocoon. The cocoon may be flat or spherical, woolly or papery, and sometimes very tough. It may be suspended in a web, enclosed in a retreat, fixed to the underside of a leaf, or buried in the soil. Some spiders guard and defend their eggs, but those that produce more than one cocoon tend to abandon them, relying on camouflage for protection.

Many mothers die before the birth of their offspring, but some others are capable of maternal care, such as wolf spiders and nursery-web spiders. However, in only a minority of cases does actual feeding of the young occur. The simplest form of feeding is the passive provision of prey. But in a few species, for example among some tangle web weavers, such as the European *Theridion impressum*, the mother regurgitates food for her spiderlings. They cluster around her mouth to imbibe a fluid mixture of predigested food and the mother's own intestinal cells. After their first moult, the young *Theridion* begin to share their mother's prey items, which she softens with punctures from her fangs. In North America and Europe, the Maternal-social Spider (*Coelotes terrestris*) takes maternal feeding to the extreme. When she dies, her tissues break down and the spiderlings feed on her body.

Above: *Red Widow Spider (*Latrodectus*) with multiple eggs.*

17

POWERS OF DISPERSAL

Much of the adventure and risk in the life of a spider occurs in the first few days of freedom when the spiderlings hatch. To avoid overcrowding and cannibalism, the young individuals disperse. They quickly colonize any new or unoccupied territory; for example, if a room is left undisturbed, spiders will move in. And they are usually among the first colonists when islands are thrown up by volcanic eruptions. Less than one year after the explosive eruption of Krakatoa in 1883, a visiting biologist on the new, barren island of Anak Krakatoa ('child' of Krakatoa) was able to find nothing alive except for one recently arrived spider. Fifty years later on, more than 90 species of spiders were recorded from the island and today this number has probably doubled.

Ballooning and the phenomenon of gossamer

Spiders populate remote places by a method of airborne migration known as ballooning. With the aid of long strands of silk drawn out by the breeze, small and young spiders, under 1cm in length, become airborne and may drift considerable distances depending on the weather conditions. They have been collected on ships, hundreds of kilometres from the nearest land, as well as in samples of aerial plankton at altitudes up to 5,000m. Charles Darwin, writing in the *Voyage of the Beagle*, recorded that on 1 November 1832, at a distance of 100km from the coast of South America, the ship's rigging became covered with vast numbers of dusky red spiders about 2mm in length. It happened again 25 days later, demonstrating that such a fall of spiders at sea is not a rare event.

At the right time and place, it is quite possible to observe what actually happens during

"THE BEST SEASON TO OBSERVE SPIDER BALLOONING IS LATE SUMMER AND AUTUMN. OVER WOODS, FIELDS, MOORS AND HEATHS, ON PLEASANT SUNNY MORNINGS, MANY SILK LINES CAN BE SEEN FLOATING ON THE AIR AND GLEAMING IN THE SUN. DURING THE AFTERNOON, WHEN THE AIR IS NO LONGER RISING, SPIDERS MAY BE FOUND LANDING, INCLUDING ON PEOPLE'S CLOTHES OR IN THEIR HAIR."

take-off. The spider climbs to the top of a fence post, or other prominent point. It turns to the wind and stands on extended legs with the spinnerets uppermost. A thread is initiated and drawn out further by the breeze. Young spiderlings can take off in the most gentle of airs but larger spiders need several loops of thread and a stronger breeze. When the pull on the line is sufficient, the spider turns, grabs the thread, releases its hold on the post, and away it goes.

Do spiders have any control over their flight? Theoretically, they do. They can descend more rapidly by rolling the thread into a ball and tucking their legs in. Also, they can fly further by building extra silk (sails) into the thread. Probably, however, they have little or no control over the direction of flight. If it happens that they land in an unsuitable place, might they decide to take off again? Yes, this certainly happens on ships at sea; Darwin observed it on the *Beagle*. In fact it has been firmly established that spiders will move on if they find themselves in a poor site.

Gossamer, the extremely light material made of silk threads, is associated with ballooning. On days when there is much ballooning activity, either taking-off or landing, there can be more than a million spiders to the acre, each one trailing a line. Lines may accumulate

Opposite: The combined lines of many tiny spiders form a sheet of gossamer over a large compost heap.

Above: *Giant Crab* spider *(Heteropoda venatoria), also known as the banana spider. The two swollen palps of the adult male are clearly visible.*

to form a silvery sheet over the land which, on a sunny dewy morning, makes a beautiful sight. But, as the morning progresses, rising air currents break up the sheet and lift the resulting pieces of gossamer, like bits of rag, into the air. While these might seem to be like 'magic carpets', they are not in fact ridden by spiders. In English, the word gossamer derives from 'goose-summer'. In France, gossamer is known as 'fils de la Vierge' (the Virgin's threads); in Germany as 'Marienfaden' (Our Lady's threads); and in Japan as 'yukimukae' (ushering in snow). Unfortunately, gossamer events these days rarely seem to reach the epic proportions of earlier times, probably because of the widespread use of pesticides.

Species that are able to disperse by ballooning tend to have a wide distribution in the world, while those that cannot balloon are limited to walking. At one end of the scale are cosmopolitan species with the ability to disperse and tolerate a wide range of habitats, while at the other end are species confined to specific niches in limited geographical ranges. For example, the Great Barrier Reef islands have been colonized by a variety of spiders (of many different families), but not by the large tarantulas which are common on the mainland, a short distance away. The young tarantulas are usually too heavy to balloon and neither can they swim. Today, many species have spread around the world thanks to international trade. One such is the so-called banana spider (*Heteropoda venatoria*). Transported together with bananas, it has now spread throughout the tropics and sub-tropics as far north as Florida and Israel. Europe, however, remains too cold for it to survive permanently.

PREY CAPTURE AND FEEDING

Spiders are usually not fussy about what they eat as long as it is captured alive. They feed mainly on insects and other spiders. However, some insects may be refused, including ants, wasps, distasteful bugs, beetles and caterpillars. Large Tarantulas will attack small vertebrates if available. In contrast to scorpions and other arachnids, spiders lack features such as grasping claws to help subdue prey. Their jaws are relatively small and their bodies are easily injured. Thus the ability of spiders to stop prey and opponents quickly, by snagging them in a web or with a venomous bite, is of critical importance (see page 86). For example, the wandering spider *Cupiennius salei* delivers a bite within 0.2 seconds of grabbing the prey

A web-building spider also works quickly because the prey could break free. If an insect contacts an orb web but then remains quiet, the spider will tweak the threads to probe its weight and exact position. Moving towards the victim, the front legs briefly touch it, while the hind legs begin to wrap silk around it. A quick bite follows. Then the neatly wrapped package is cut out and carried to the web's centre (hub) or to the spider's hiding place.

The net-like orb webs would appear to be the most efficient means of intercepting plenty of insects. However, many hunting spiders, without the use of webs, catch insects two to three times their own size. For example, the relatively small crab spiders wait among flowers and grab large bumble-

Below: Nocturnal wandering spider (Cupiennius getazi), a native of Central America.

bees, arresting them quickly with a potent bite. In some species the diet is highly specialized, as in the American Bolas Spider (*Mastophora bisaccata*) which catches males of a single species of moth. At the other extreme, the Mouse Spider (*Scotophaeus blackwalli*), common in buildings in many countries, will eat any small insect including those that are already dead; it will even eat an insect on a pin in a museum collection.

Spiders cannot swallow solid food and so digestive juices are regurgitated to liquefy the inside of the prey. The stomach works like a pump to suck in the resulting soup. Some, such as the tangle web-weavers and the crab spiders, leave their prey looking almost intact after sucking out the tissues, because their jaws, lacking teeth, are not particularly strong. On the other hand, the typical orb web-weavers, with teeth on their jaws, mash the prey to a pulp.

When prey is abundant, the abdomen swells to absorb the food. However, when times are hard, many spiders, because of their low metabolic rate, can endure months without feeding. And many have no need to drink water directly. The lowest metabolic rate of all spiders occurs in those living in caves. A species such as the European Cave Spider (*Meta menardi*) consumes only one-eighth the oxygen of a similarly-sized Garden Spider (*Araneus diadematus*).

Right: *Green Lynx* (Peucetia) *with prey, on cactus.*

DEFENCE AGAINST PREDATORS AND PARASITES

Spiders are fairly soft bodied and not distasteful. Therefore they have many enemies such as birds, reptiles, amphibians (especially toads), mammals, ants, wasps, preying mantises, scorpions, centipedes, and other spiders. Certain bats can pick up spiders from the ground and out of their webs. When threatened, some hunting spiders quickly adopt defensive postures. One of the best examples is the Brazilian Wandering Spider (*Phoneutria nigriventer*). Standing on its back legs, the front legs rise and the jaws open. Tarantulas also adopt this position and may keep it up for several minutes.

Species such as the Green Lynx Spider (*Peucetia viridans*) of North and Central America defend themselves by spitting venom in the direction of the enemy. The spray can reach 20cm and is stinging if it gets in the eyes. But the great majority of spiders rely simply on their camouflage for protection. For example, spiders coloured green live among leaves, while red, yellow and white species live among flowers. Others have colours that blend well with backgrounds such as sand or lichen. Some species are virtually impossible to see because of their cryptic body posture. They may look like a twig or even a bird's dropping. Some nests may be wonderfully concealed, such as those of the cosmopolitan sac spiders (*Clubiona* species), which fold over a leaf and build their nest within.

Some spiders mimic ants, since predators avoid ants with their painful stings. Some of the best examples are among the jumping spiders, for example *Myrmarachne* species. They are often unrecognized as spiders; the difference is that they have four pairs of legs instead of the ant's three. But when a column of voracious army ants approaches, many spiders use a simple defence. They drop from a branch and dangle by a thread (dragline) until the ants have passed. A few, such as the jumping spider *Pellenes nigrocil-*

iatus, of Europe and Asia, go as far as to suspend an empty snail shell on a line above the ground as a shelter. However, the labour necessary to haul up such a shell must be enormous considering its weight is likely to be five or more times that of the spider.

Spiders also have enemies within. For example, pirate spiders are small, delicate species that have the temerity to attack other spiders in their webs. They use a form of trickery known as aggressive mimicry. *Ero cambridgei*, for example, imitates the courtship signals (thread plucking) of the male Autumn Spider *Metellina segmentata* of Europe. The female *Metellina* makes the mistake of moving towards the source of the vibes and falls victim to *Ero* which has a rapidly acting venom. Less menacing invaders include the parasitic spiders (*Argyrodes* species), known as kleptoparasites, which simply steal food from the webs of larger species.

Above: *Tachina fly larvae eating Tarantula spider.*

Spider-hunting wasps

When confronted with their arch enemy, the spider-hunting wasp (Pompilidae), many spiders seem to be almost defenceless. The largest spider-hunting wasps, the Tarantula Hawk Wasps, are the most impressive wasps in the world, having a wingspan of up to 10cm. Some observers have described epic battles which ended with the tarantula paralysed by the wasp's sting, while the wasp itself died from poisonous wounds made by the fangs of the tarantula. But others have found the tarantulas to be at a great disadvantage. Handicapped by almost total blindness, they have been forced to meet one surprise attack after another, with fangs as weapons that are utterly unsuited against the enemy.

Some spider-hunting wasps search for trapdoor spiders living on sand dunes. They dig into sand nearby to break into the spider's burrow and drive the owner out. When the spider emerges, the wasp attacks and paralyses its victim, drags it back to the burrow, deposits a single egg on its abdomen and fastens the trapdoor. When the egg hatches, the larva feeds on the living tissue of the spider. Another group of wasps, the mud dauber or digger wasps (Sphecidae), target other spiders, including web-builders. They may take and paralyse as many as twenty specimens in a day to provision their nests. The wasp grubs hatch in the nest and feed on the stored food until they pupate.

From the moment it is stung by a species of ichneumonid wasp, the brightly coloured orchard spider *Leucauge argyra* of Central America, is doomed. The egg that the wasp has laid becomes a growing larva sucking fluid from the spider's abdomen. For a couple of weeks, the spider continues with its daily routine of building a web and catching insects. But then, the parasite takes control and injects a chemical into the spider which programmes its brain to build a unique 'cocoon web'. This web is useless to the spider but perfect as a structure to protect the cocoon of the wasp larva during its last weeks of development. It is the spider's final work.

Opposite: Orchard Spider. A beautiful orb-weaver from tropical regions.

Below: Spider-hunting wasp taking paralysed spider prey to its nest.

CHAPTER 2
SPIDERS THAT HUNT

There are many varieties of hunting spider belonging to a large number of families; these include the wolf spiders, wandering spiders, nursery-web spiders, fishing spiders, huntsman spiders, jumping spiders and lynx spiders. They do not build webs but employ various tactics to catch prey. Many use their camouflage combined with the element of surprise; others use speed and strength to overcome their victims. Hunting spiders generally have the best eyesight. Often the females carry their egg cocoon with them and some also carry the babies for a time.

SPIDERS DIVIDE INFORMALLY INTO (A) HUNTERS AND (B) WEB BUILDERS. HUNTING SPIDERS FORM AN EXTREMELY DIVERSE DIVISION. SOME ARE HIGHLY ACTIVE AND OTHERS ARE MORE SEDENTARY; THE ONE THING THEY HAVE IN COMMON IS THAT THEY DO NOT BUILD WEBS. A NOTABLE EXAMPLE OF A SEDENTARY HUNTER IS A SOUTH EAST ASIAN CRAB SPIDER, 6MM IN LENGTH, CALLED THE PITCHER-PLANT SPIDER (MISUMENOPS NEPENTHICOLA). THE CARNIVOROUS PITCHERS COLLECT WATER AND LURE INSECTS WITH THEIR SUGARY SECRETIONS; ANY INSECTS THAT DROWN ARE ABSORBED BY THE PLANT. THE SPIDER LIVES UNDER THE CIRCULAR RIM OF THE PITCHER AND USES THE HABITAT TO FIND FOOD AND ALSO PROTECTION FROM PREDATORS. INSECTS MAY FALL INTO THE WATER BUT NOT THE SPIDER! IT FISHES FOR CASUALTIES AND IS ABLE TO CATCH PREY SUCH AS ANTS WHICH WOULD OTHERWISE BE TOO DANGEROUS.

Wolf spiders

Previous: Huntsman spider (Sparassidae *sp.*).

Below: Wolf spider with prey.

Wolf spiders are so called because people wrongly imagined that they hunted in packs like wolves. In fact, they are independent hunters. Those without homes or burrows are among the most active of spiders, but those with burrows tend to be sit-and-wait predators. They are found in most parts of the world and are abundant in many places. If you see numbers of spiders speeding away on the ground and several have white or blue egg sacs attached to the abdomen, you can be sure they are wolf spiders. Often the egg-sac is bigger than the abdomen

itself. For several weeks, the females seek patches of sunlight to warm their eggs. The most common wolf spiders, the relatively small *Pardosa* species, are most active when the sun is shining, and in poor weather they hide among leaves and detritus.

While some wolf spiders are active during the day, others are nocturnal. Most live on the ground and many of the larger species dig burrows from which they emerge to hunt. In places, the burrows can be easy to spot because they have a woven 'turret' around the entrance. Species such as *Lycosa ingens*, of Madeira, are among the largest of the araneomorphs or true spiders (with a body-length up to 38mm). Most are sombre-coloured but many are quite handsome in appearance. The area around the mouth-parts may be coloured yellow or orange. Wolf spiders have relatively large eyes, enabling them to locate prey at a distance of up to 10cm. Of their eight eyes, the large principal pair look forward and one other pair look upwards. These spiders attack their prey vigorously, crushing it with stout jaws.

Female wolf spiders may be tempera-mental but they are also known for their maternal care. In common with other wolf spiders, the European *Lycosa narbonensis* protects her egg-sac by carrying it every-where, attached to the spinnerets. She has a powerful instinct to defend it but she can be easily fooled: if her egg-sac is changed for something artificial, like a piece of cork or a wad of paper, she will defend the substitute with her life. After two to three weeks, the mother bites open the sac to allow the brood of up to 100 spiderlings to climb onto her abdomen, several layers deep. Living on their reserves, they hold on for about a week while she continues to hunt, and defend herself if necessary. However, there is no mutual recognition. Females accept spider-lings from another female, and the spiderlings will climb onto the backs of other spiders, even males of other species, which often simply eat them.

Some other types of wolf spider, for example the European *Arctosa perita* and *Trochosa ruricola*, do not carry the cocoons with them but keep their eggs cool under the ground in silk-lined holes or tubes. Also, in Africa, there is a minority of wolf spiders, called wolf weavers (*Hippasa* species), that differ from the rest in constructing a sheet-like web. Some wolf spiders live close to water and can even walk on its surface. Experiments conducted on the shores of lakes in Finland have shown that when spec-imens of the species *Arctosa perita* are placed out in the water they hurry straight back to the bank. If they are taken to the

Above *Wolf spider (Lycosid) with orange jaws.*

Above: *Fishing spider (Dolomedes minor) on cornflower.*

habits and maternal care. The egg-sac, or cocoon, instead of being attached at the back of the body, is carried by the female under her body. It is held in her jaws and further secured by silk lines to the spinnerets. It is often so large that her legs can barely touch the ground. Near the time for hatching, she fixes the cocoon to the vegetation and surrounds it with a network of threads to form a web-like tent. It is not a trap for catching prey, but a protected space for the spiderlings. When they are ready to hatch, the cocoon is opened by the mother. The youngsters remain in their nursery until their first moult. In North America, Europe and Japan, the mother on guard beside the nursery-web is a common sight in rough grass during the summer.

The semi-aquatic, fishing or 'raft' spiders, such as *Dolomedes fimbriatus* in Europe and *Dolomedes triton* in North America, are found in many parts of the world. They are quite large spiders (females can reach a body length of 30mm), living and hunting on the surface of still or slow-moving freshwater. They row themselves across the surface, like pond skaters (Gerridae), but they also sail across, taking advantage of the wind. Their legs, which radiate out on the surface, are highly sensitive to vibrations from floating insects. They may also put one leg in the water to attract a fish. If a small fish passes under its outstretched legs, the spider makes a swift and sudden plunge, breaking through the surface tension. The fish may be up to four times the weight of the spider, but it is seized by the legs and bitten by the powerful jaws. The prey is dragged across the water and eaten at the bankside. Like the nursery-web spiders, female fishing spiders hold their large egg cocoons in their jaws. Their main enemies are predatory fish, such as trout, but *Dolomedes* will try to escape by jumping vertically upwards.

A species of fishing spider in South America (*Diapontia uruguayensis*) has great skillfulness as a fisherman for tadpoles. On the edge of the water, often between stones,

opposite shore and placed in the water there, they attempt to cross to the shore on which they live, but only if the sun is shining. They navigate according to the position of the sun.

Nursery-web and fishing spiders

Nursery-web spiders (*Pisaura mirabilis*) resemble wolf spiders except in their mating where the tadpoles like to sun themselves, the spider constructs a funnel-shaped net, a portion of which dips into the water. The tadpoles, without suspecting the cunning of the spider, venture into the funnel, whereupon the spider rows across to drive them further in and then attacks any that have ventured too far inside.

Huntsman spiders

Huntsman spiders are fast-moving and crab-like. They tend to be large and rather flat, with legs extending outwards but curving forwards. Their distribution is mostly tropical and subtropical. The family (Sparassidae) includes the cosmopolitan banana spider (*Heteropoda venatoria*), which often travels throughout the world in banana shipments. In Europe, there are only a few representatives of the family, of which the most notable, with its bright green colour, is the Meadow Huntsman *Micrommata virescens*. The adult male is particularly pretty with green, yellow and red colouring.

The Lichen Huntsman (*Pandercetes* species) of forests in Southeast Asia and Australia is a beautifully camouflaged spider. It lives on mossy, lichen-covered tree trunks and matches the colour and texture perfectly, making it very difficult to see. If approached too closely, it moves at great speed to the other side of the tree. The hairiness of the legs and body, pressed close against the bark, helps to blur the spider's outline and reduce shadows.

Spectacular examples of huntsmen include the Golden-wheeling Spider (*Carparachne aureoflava*) and the Dancing White Lady Spider (*Leucorchestris arenicola*), both from the Namib Desert dunes, southern Africa. At night, these spiders wander over a large area of the dunes in search of mates and prey but always return to their burrow. *Carparachne* has a unique method of fleeing from its main enemy, the spider-hunting wasp. It throws itself sideways and cartwheels down the dune. In other words, it

invented the wheel before humans! It achieves twenty revolutions per second and travels downhill at a speed of one metre per second.

The wandering spider (*Cupiennius salei*) of Central America belongs to the family Ctenidae but resembles the huntsman spiders. It spends most of its time lying in ambush. It is extremely patient and highly sensitive to vibrations across the substrate or through the air; it does not depend on its eyesight. An attack is launched when prey comes close, and is extremely fast. The spider grabs the prey, makes a judgement

Below: Meadow Huntsman spider (Micrommata virescens).

Right: *Yellow crab spider with prey larger than itself.*

Opposite: *Running crab spider (philodromid) in resting pose.*

through its tactile sense of the prey's suitability, and bites it, injecting venom, in less than 0.2 seconds.

Crab spiders and running crab spiders

Crab spiders form a large family (Thomisidae) of fairly small spiders which typically have squat bodies. The European *Xysticus cristatus* is a good example. It can walk forwards, backwards and sideways. Crab spiders do not spin webs; they are lie-in-wait ambushers which rely heavily on camouflage. The first two pairs of legs, used for grasping prey, are longer and stronger than the last two pairs. Mostly they are not very hairy. Some can change their colour to match the background, for example the European *Misumena vatia* (female). The colour change process takes about four days. Many crab spiders sit in flowers using the colour, scent and nectar to lure butterflies, bees and other insects within easy reach. Often the prey is larger than the spider itself. For example, a stout bumble bee is quickly overpowered from a bite with a highly potent venom. A crab spider will kill its prey and suck it dry through the tiny bite holes. It possesses no teeth on its chelicerae, so that the prey looks untouched; a bee sitting motionless on a flower is often the sign of a crab spider. The crab spider may remain hunting for days, or even weeks, at the same spot. Unfortunately, this can be its downfall as it may then be found by a predator. Males are invariably smaller than females and quite different in appearance. The egg sacs are fastened to the vegetation and are often flat in shape; the females usually stand guard over the cocoon.

The Heather Spider, *Thomisus onustus*, found from Europe to Japan, may be pink, yellow or white, depending on the colour of the flower on which it is sitting; in heather, for example, the spider is pink. Insect prey may not notice it and predators miss it as well. The female has a broad, angular abdomen and the carapace has two horn-like projections near the eyes; the smaller male is darker. *Synaema globosum* is a small, rounded crab spider common in the Mediterranean region where it can be seen in flowers such as umbellifers. The colour of this spider is black with a red, yellow or white pattern.

A particularly handsome species is the White Crab Spider, *Thomisus spectabilis*, of forests and gardens in Southeast Asia and Australia. The legs and carapace are translucent white, while the abdomen is pure white. The little brown male is so different in appearance, he seems to be unrelated. The Seven-spined Crab Spider, *Epicadus heterogaster*, of South America is another beautiful species, but a strange one, which appears to imitate a white, orchid-like flower. Seven large tubercles arise from the abdomen, which is pearly-white, and the carapace and legs are translucent white. Insects which visit the 'flower' to sample its nectar unexpectedly find themselves in the deadly embrace of the spider.

Above: *Regal Jumping Spider. An exceptionally hairy spider.*

Jumping spiders

Jumping spiders are everyone's favourite. Their compact bodies, large eyes, high levels of alertness and bright, iridescent colours give them a charming appearance. If a person approaches a jumping spider, it will often turn and stare back. And photographers may discover that their subject has jumped out of the view-finder and on to the lens! In fact, it can be great fun to offer a small mirror to a male jumping spider, which is likely to jump at the reflected image, thinking it is a competitor. No other spider will do this.

Jumping spiders represent the largest of all families with over 4,000 known species in the world. Because of their exceptional eyesight and other abilities, they are considered to be the most advanced and highly evolved family. Their scientific name, the Salticidae, is derived from the Latin *saltus*, meaning 'a leap'. In spite of being fairly small, with a body length rarely more than 15mm, they are formidable predators of insects. They stalk and pursue their prey until it is close enough for a final pounce. Salticids are usually active during the daytime, especially in sunshine. They occur in most places but are particularly abundant in tropical regions. They almost never spin webs but use silk for a variety of other purposes.

The ability to jump up to twenty times its own length requires good eyesight. The arrangement of the eight eyes on the head of a jumping spider gives it almost 180-degree vision. The central, forward-looking pair (principal pair) are large and can recognize shapes, colour, and distances up to 20cm. Despite the tiny dimensions, the eyes are based on long tubes, which work like miniature telephoto systems. They may be called simple ocelli but in fact they are highly developed. Having recognized its prey, and stalked it until sufficiently close, a jumping spider will fasten a safety line (dragline) to the surface before using its four back legs to launch it forwards. It then sails through the air with its relatively stout front legs held out to grasp the prey.

Looking like a bird dropping is a useful disguise for some spiders. The Bird-dropping Spider *Phrynarachne rugosa* of Africa is quite bizarre. The blobs and warts all over its dark and pale, glazed surface give the spider a wet and lumpy look. It draws in its legs and waits motionless on a leaf for hours and hours. Sometimes it adds to the deception by sitting beside a small, messy-looking patch of white silk. Also, cunningly, the spider emits a smell to attract flies, which are lured straight to its grasping legs.

Not all crab spiders have a chunky build. Other species, such as *Tibellus oblongus*, have long slim bodies and lie, unseen, along plant stems. In the running crab spiders (Philodromidae), in which the legs are of equal length and the sexes are relatively similar, prey is usually caught by speed, as with many other hunting spiders. Most running crab spiders are active on vegetation. However, the unusual species *Philodromus dispar* is an inhabitant of buildings in Europe, much of Asia and North America. The male is quite distinctive, with a body appearing iridescent blue-black.

The Zebra Spider, *Salticus scenicus*, common in Europe and North America, can often be seen on walls, where it appears to defy gravity, like Spiderman! It jumps off and, instead of falling down, swings back to safety by virtue of its dragline.

Jumping spiders are known for their elaborate courtship behaviour. The males dance in front of the females, waving their legs, excitedly drumming their palps and adopting poses to show off their colours. The sexes differ in appearance and males are often distinguished by their larger jaws, which are used to restrain mating partners and competing males. Many salticids make a silken sac-like shelter in crevices, under stones or bark, or on foliage. They retreat to these sacs at night, in rainy weather or during the winter to hibernate. The sacs are also used to protect the eggs.

The Jumping Spider *Portia*

One of the most extraordinary jumping spiders is *Portia fimbriata* of tropical Australia. It is not pretty or colourful but, because of its wide repertoire of hunting tactics, it seems to be the cleverest of all spiders. It walks slowly in a robot-like fashion and looks nothing like a normal spider. Even other jumpers do not recognize its stealthy approach. And, when in a web, it is easily mistaken for a piece of detritus. The female measures about 1cm in body length and the male is a little smaller.

One of the tactics of *Portia* is to invade the webs of other spiders – which is most unusual. Typical hunting spiders usually move with considerable difficulty in webs; and web-building spiders are ill at ease in the webs of other kinds. But not *Portia*. It can move about and capture prey in all sorts of webs. It can also spin its own web, which is quite unusual for a jumping spider. Sometimes *Portia* builds its web adjacent to another spider, so that when that spider follows an insect in hot pursuit across the web of *Portia*, it can be attacked. Another trick is to make vibrations to deceive the neighbour. If the spider next door emerges, expecting an insect, *Portia* leaps onto it. This spider's jumps are quite

Above: *Jumping spider* Portia. *With its huge eyes and palps resembling boxing gloves, the appearance of this spider is quite bizarre.*

Right: Jumping spider. The time-lapse photography shows take-off, followed by flying through the air trailing a dragline, and landing.

Above: *Green Lynx on flower.*

impressive. It can jump, directly upwards, as much as 10cm. Even its tactics on landing are quite devious. *Portia* either freezes or runs about 10cm and then freezes.

Predation on other spiders is a dangerous occupation but *Portia* has a secret weapon – its exceptional vision. It can distinguish mates and prey at distances of up to 27cm, which is a 7cm advantage over other jumpers. Not only is *Portia* the sharpest-eyed of all spiders, but its optical resolution is said to be superior to all other terrestrial inverte-

brates, most of which have compound eyes. The principal eyes of *Portia* are of the 'simple' type and comparable with our own, but what is most remarkable is the tiny space they occupy. The size of the retinal receptors is close to the theoretical minimum, given the physical properties of light.

Lynx spiders

Whereas wolf spiders hunt on the ground, lynx spiders (Oxyopidae) tend to hunt on plants. They are agile and capable of running and jumping on their prey like a cat, hence the name lynx spiders. They occur in Europe and North America but are most common in tropical regions. The family includes golden lynx spiders such as the European *Oxyopes lineatus*, and green lynx spiders such as the North and Central American *Peucetia viridans*. The latter are particularly pretty. Lynx spiders have a noticeably pointed abdomen, a dome-shaped cephalothorax and legs with numerous spines standing out at right angles. They can reach 12mm in body length. Although their eyesight is not as good as that of the jumping spiders they can see their prey from a distance of up to 10cm. Female lynx spiders do not carry their egg-sacs with them but attach them, with a mesh of silk threads, to a plant and stand guard over them. Being unable to hunt and stand guard at the same time, however, the mother eventually dies.

Night crawlers and sac spiders

Night crawlers (Gnaphosidae) are nocturnal hunters. They are often dark-coloured and their eyes include a silvery oval pair. Their distribution is worldwide. The mouse spider (*Scotophaeus blackwalli*) is a common example. It has a grey, furry body which is low and flat. In Europe it is a nocturnal wanderer on walls, where it searches for insect prey. Sac spiders (Clubionidae) are similar to night crawlers but usually paler in colour. They are common in most places on vegetation and some species also occur in buildings. The posterior median eyes are circular, rather than oval as in the night crawlers. The typical sac spider, for example the European *Clubiona terrestris*, is a relatively small, pale species with a glossy brown carapace and a silky grey, elongate abdomen. Males and females are quite similar. They usually make a sac-like retreat in rolled leaves, or folded blades of grass, or under

Below: *Mouse spider viewed from below. The pale squares of the two book lungs are visible at the front of the abdomen.*

Above: *Ant mimic spider.*

loose bark. The retreat is used for egg-laying, moulting and resting, the spider emerging at night to hunt insect prey. The slender sac spiders, such as *Cheiracanthium mildei* of Europe and North America, are recognized by their long first pair of legs. The genus is known for the fact that some of its members are venomous.

Not all species of night crawlers are nocturnal. The strikingly-marked Vespa Sac Spider (*Supunna picta*), from Eucalyptus forests in Eastern Australia, runs about on sunny days, waving in the air its orange first pair of legs. It runs very fast in short bursts with frequent changes of direction, appearing to imitate a solitary hunting wasp waving its antennae.

Ant-mimicking spiders

Some spiders can easily be mistaken for ants. The Trinidad ant spider (*Myrmecium*

species) is a night crawler that has become a remarkable mimic of a red ant. The palps of the male spider look like the jaws of the ant and the front legs are waved like antennae. The carapace is shaped to look like the separate head and thorax of the ant. The illusion is completed by the false waist and the small, glossy brown abdomen. If this spider is compared with the Spear-jawed Jumping Spider (*Myrmarachne plateleoides*) of Southeast Asia, which itself is a perfect mimic of the Red Weaver Ant (*Oecophylla smaragdina*), the similarity of the two spiders is quite striking. *Myrmarachne* has ant-like movements and it also raises its front legs like antennae. Its deception is sufficiently successful to enable it to enter the ant's nest, but it feeds only on the larvae and pupae. Ant mimics also occur among the crab spiders. For example the black ant spider of Peru

(*Aphantochilus* species) makes a close imitation of a black ant (*Cephalotes* species). Either side of the spider's carapace are a pair of horns, similar to those of the ant. The eyes are widely spaced and the short abdomen is rounded, like the ant's.

Spiders living in water

The Water Spider, *Argyroneta aquatica*, is the only species that lives permanently under water. It occurs in lakes and ponds in Europe and Asia. With silk threads, it constructs a 'diving bell', attached to a water plant, which is stocked with air taken down in bubbles. The bubbles adhere to the spider's hairy body, making *Argyroneta* look silvery. But, because of its great buoyancy, it has to swim extremely energetically. Prey, including fly larvae, small crustaceans and small fish, is caught under water. When at rest, in the diving bell, the Water Spider breathes normally, as if on land. Remarkably, this terrestrial form of life has colonized the water environment with such evolutionary speed that most of its adaptations are behavioural. Apart from its presumed ability to breathe oxygen through its skin, it has no obvious anatomical adaptations to aid it in its new life.

Besides *Argyroneta*, some normally terrestrial spiders, such as wolf spiders and money spiders, are capable of submerging underwater, in freshwater, for periods as long as three weeks. They may breathe from a bubble of air surrounding the body where probably they also absorb oxygen directly from the water, through the skin. The sea, as opposed

Left: *Water spider (male). In this view of the spider as it rests upside down, the large jaws (chelicerae) are clearly visible.*

to fresh water, has proved to be a more challenging environment. Few spiders, or insects for that matter, have colonized the marine environment. Only the remarkable semi-marine spiders, such as *Desis marina*, are able to live regularly in rock pools. They occur on mostly tropical and subtropical shores, and can survive for days under the seawater, without any bubble of air and no diving bell. How they manage to do so is of great interest. Indeed, many aspects of the life of *Desis* have still to be explained. How, for example, does itsbody tolerate the saline water without any obvious adaptations?

Spitting spiders

The slow moving spitting spiders (Scytodidae) catch their prey in a way that is all their own. They have adopted the vulgar habit of spit-

ting. Species such as the European *Scytodes thoracica* spit a gooey mixture of venom and glue at prey which passes within 2cm. Their venom glands have become so enlarged, to produce both venom and glue, that the carapace is highly domed to accommodate the glands. The sticky venom is squirted through the chelicerae to paralyse the prey and fix it to the substrate. As the spittle sets, the spider comes forward and stabs the victim with its fangs, wraps it up in silk, and either eats it on the spot or takes it back to the spiderlings in the nest. Female spitting spiders have a high level of maternal care. They carry the egg-sac in their jaws but put it down while they capture and eat prey; when finished, they place the sac back in their jaws. They are nocturnal, six-eyed spiders and their distribution is cosmopolitan except in cold regions.

Opposite: *Water spider (female) with air bubble surrounding abdomen.*

Below: *Spitting spider. Dark spots are distributed over the legs and body, including the chelicerae.*

CHAPTER 3
SPIDERS THAT BUILD WEBS

Spiders' webs come in all shapes and sizes and many still await description. Some are designed to ensnare crawling insects, others to catch flying insects, while some target jumping insects such as grasshoppers. Some operate during the daytime, others work both day and night. Those that specialize in catching moths are built each evening and removed at dawn. Some are sticky to the touch, others dry, and yet others woolly. Some are extensive, some are built gregariously, and others are no more than a solitary thread. The more primitive kinds are little more advanced than the simple affairs of trapdoor spiders.

IN NORTHERN CLIMATES, AUTUMN IS THE SEASON WHEN WEBS APPEAR TO BE MOST ABUNDANT. THIS IS BECAUSE THE MAJORITY OF SPIDERS, WHICH HATCHED DURING THE SPRING, HAVE BECOME FULLY GROWN AND ARE BUILDING FULL-SIZED WEBS. A BASIC DIVISION EXISTS BETWEEN WEBS THAT ARE RENEWED DAILY (ORB WEBS) AND THOSE THAT ARE A LIFETIME'S WORK FOR THE OWNER, SUCH AS SHEET WEBS.

Observers may wonder how spiders avoid getting stuck in their own web. Essentially, they avoid walking on the sticky parts, but they do get stuck if picked up and thrown back into their web (not recommended!). This chapter is about araneomorphs (typical or true spiders) though many mygalomorphs (tarantulas) also build webs.

How webs have developed

Because of their extreme fragility, spiders' webs have not survived as fossils, and thus we can only theorize about their evolution, using the webs in existence today. However, the interesting thing is that, when you look at a web, you see a fixed record of animal behaviour. Describing and classifying this behaviour helps us to try to reconstruct the story of web evolution. Clearly an evolutionary force, or a kind of 'arms race', has always existed between spiders and insects. For example, insects evolved the power of flight to escape spiders on the ground, but spiders responded by evolving aerial webs to catch the flying insects.

It is probable that the earliest spiders' webs were little more than extensions of silk from a hiding place. As spiders darted out, they trailed their draglines behind them. At first, the lines acted as simple trip-lines to signal the approach of prey. Later, the webs diversified: some developed into horizontal sheets, some became trampoline-like structures, and finally, according to the theory, some webs turned upright and became two-dimensional nets with the spider sitting at the centre.

Primitive types of web

The kinds of web considered to be most primitive are the simple

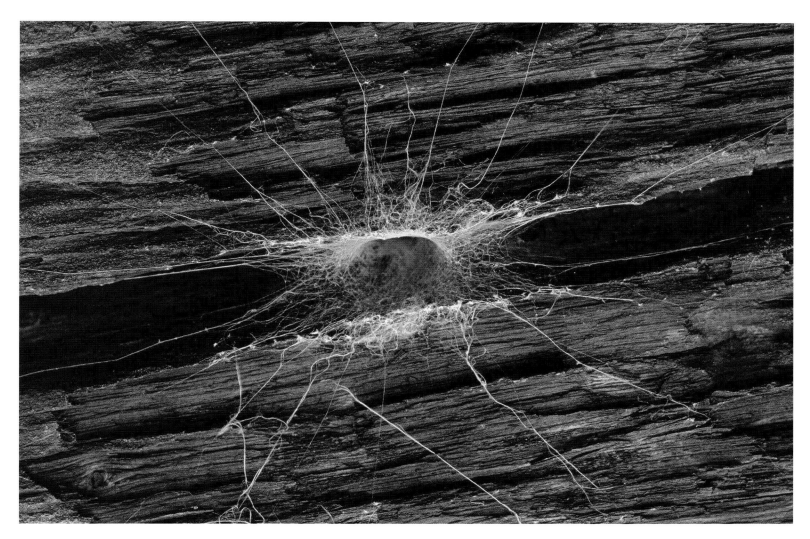

tube webs that radiate signal threads from a silk-lined hole or burrow. Such a web does not actually capture prey but only transmits vibrations from a passing insect. The spider rushes out from its retreat and grabs it. This kind of web appears to have been used by many of the earliest spiders, 300 to 400 million years ago. Today, this type of web continues to be made by a number of species, including the European Tube Web Spider (*Segestria florentina*). It is also not unlike those still made by the 'living fossils', the giant trapdoor spiders (liphistiomorphs) of Southeast Asia (see page 74).

The Tube Web Spider has spread from Mediterranean Europe to many other temperate parts of the world, where it is usually found on walls and fences. It is a six-eyed spider with an elongated body, reaching 22mm, that fits easily within a hole or crevice. As twilight comes, the spider appears at the entrance to its retreat and maintains leg contact with each of the radial

threads. The wait for prey, however, may be a long one as walls are relatively sterile habitats. Nevertheless, *Segestria* instantly emerges to attack anything, at any time of the day or night, that touches the threads around its retreat. Prey includes flies alighting to bask in the sun, other spiders, beetles, bees and centipedes. The one season of plenty for the Tube Web Spider is late summer, when there is a mass emergence of those gangling insects, the Crane-flies.

A type of web which has a proper catching area, but remains relatively unsophisticated, is the flimsy, vaguely funnel-shaped web, almost invisible, that is used by the six-eyed, Daddy-long-legs Spider *Pholcus phalangioides*. This pale grey, fragile-looking species hangs upside down in its web in dark and damp places, such as cellars. However, it is also common in buildings generally. It is a prolific breeder and is spreading in many countries. When an insect gets tangled up in the web, the Daddy-long-legs

Above *Tube web belonging to* Segestria florentina.

Opposite: *Orb-weaving spider with rudiments of a web.*

Previous: *Orb web spider on web.*

47

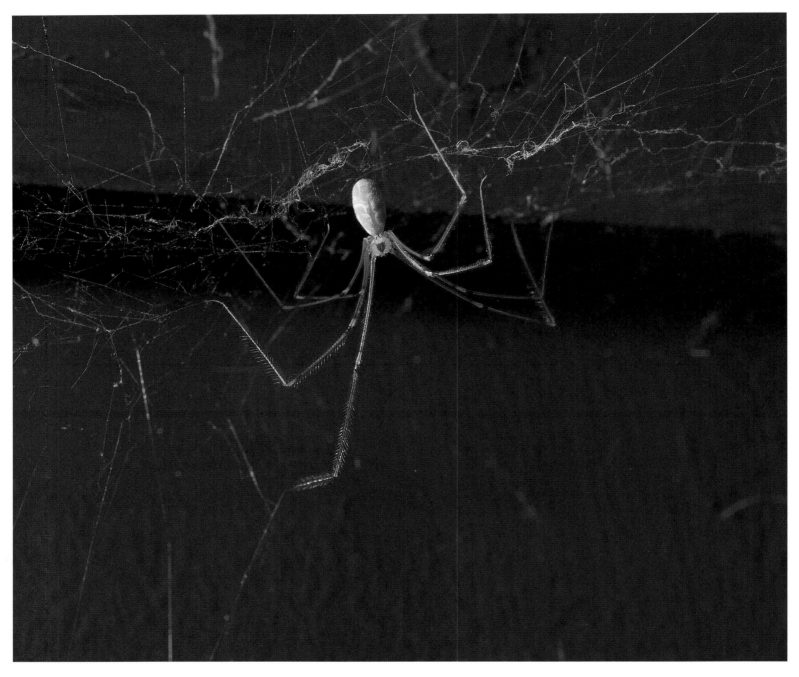

Above: *Daddy-long-legs Spider* Pholcus phalangioides. *This is a spider that responds to disturbance by bouncing in its web to become a blur.*

uses its long legs to throw fresh strands of silk at the prey from a safe distance. It also invades the webs of other spiders: among its prey are the much larger and fiercer house spiders of the genus *Tegenaria*. The mother spider carries her loosely wrapped eggs, and later the babies, in her mouthparts.

When disturbed in its web, the Daddy-long-legs rapidly bounces up and down to scare or distract intruders. It will also tug, or bounce, to discover the weight of objects in the web, such as insects playing dead. When the spiders live in a dense colony, this bouncing can amount to a communication system. As an experiment, it can be interesting to make a small disturbance in the web of

a Daddy-long-legs, and to see if a reaction occurs among any of the neighbours. How far does the information travel?

The six-eyed recluse spiders, for example *Loxosceles reclusa* of North America, are somewhat similar to the Daddy-long-legs spider. They also spin flimsy, untidy webs in dark spaces under rocks, in holes in the ground, or in dark corners of buildings. They are brown in colour, small- to medium-sized spiders with fairly long legs. They are notorious, particularly in North America, for being venomous to man. The species *Loxosceles rufescens* is virtually cosmopolitan, while other species occur in America, Europe and Africa.

Tangle webs

A considerably more complex and sophisticated type of web, with special catching threads, is built by the tangle web-weavers (Theridiidae). The False Widow Spider, *Steatoda paykulliana* of the Mediterranean region, is a good example. Tangle webs are three-dimensional structures that occupy spaces in dark corners or under shrubs. Many have a built-in, thimble-shaped retreat in which the round-bodied spider waits. Such webs are somewhat irregular but three structural levels can be recognized: an uppermost trellis of supporting threads, a central zone of tangle threads, and a lower zone of vertical catching threads. Under tension, the catching threads, which are quite ingenious, are beaded with sticky droplets near their attachment to the substrate. Passing insects which break a thread's attachment find themselves stuck to the gum and lifted towards the central tangle as the thread contracts. Struggling only causes further entanglement and, after throwing more gummy silk over the victim, the spider finishes it off with a bite. Although relatively small spiders, they are not afraid to attack large insects and may even hoist a lizard into their web. Tangle webs can also catch ants, whereas orb webs are unlikely to do so.

Above *Tangle web. The owner of this web is small and difficult to see inside the central tangle.*

Above: Sheet web. The spider, a member of the family Linyphiidae, runs upside-down on the underside of its sheet.

Sheet webs

An evolutionary advance beyond the tangle webs is demonstrated by the sheet webs. In these more or less horizontal webs, the central tangle has become a distinct sheet. The weavers of these webs represent a group of about a dozen families ranging in size from the small money spiders (Linyphiidae) to the medium-sized sheet web-weavers (Agelenidae). Some of the most familiar kinds of sheet webs belong to the *Tegenaria* house spiders, such as the European *Tegenaria parietina*, whose dusty drapes are so much a feature of unswept outbuildings and garages in many parts of the world. Sheet webs built

indoors are sometimes called cobwebs and those with a tubular retreat at one end tend to be called funnel-webs.

Some spiders, such as the Labyrinth Spider (*Agelena labyrinthica* in Europe or *Agelenopsis aperta* in America) run on top of their sheet. Their broad, slightly concave sheets cover grasses or shrub vegetation and usually funnel at the back into a retreat. They have no sticky silk but use a mass of criss-crossing lines above the sheet to knock down flying insects. These fall and stumble about, completely out of their depth, in the maze of trip-lines. The arachnologist W.S. Bristowe compared such an insect to a man trying to run through soft snow up to his knees, while pursued by an enemy on skis. In fact it is amazing how fast the spider darts across its web, given that it has eight legs to control. The typical sheet web is permanent and enlarged as the spider grows. Feeding and egg-laying take place in the funnel and the male often stays with the female prior to mating (co-habitation).

Other sheet web-weavers, for example the money spiders or linyphiids, hang upside down below their sheet, as in the European and North American *Linyphia triangularis*. Their three-dimensional, slightly convex, hammock-like webs show up very well in the early morning dew. Fifty or more such webs may be found on a single shrub, if it has the firm branches favoured by spiders. Flying insects hit the web's upper scaffolding and are knocked onto the sheet. *Linyphia* shakes the web's structure to assist the insect's descent, and then bites through the sheet to pull the victim through. The damage is repaired later. Linyphiid spiders are abundant in temperate regions and are the dominant balloonists. Males are not much smaller than females.

Orb webs

When horizontal sheets were able to transform into vertical sheets, probably earlier than 110 million years ago, the first orb webs had arrived. They became the most economical and effective trap for aerial prey. Today, there are around 3,000 species worldwide that build the two-dimensional structures resembling a wheel with spokes, which we call orb webs. Essentially, orb webs consist of three elements: strong support threads that form the framework, strong radial threads which converge at the centre, and an elastic capture thread which spirals round and round and is beaded with sticky droplets.

The main families of orb-weavers are the Araneidae (typical orb-weavers), the Tetragnathidae (long-jawed orb-weavers) and the Uloboridae (cribellate orb-weavers). Their

handiwork is often noticed because the spiders and their orb-webs are sometimes quite large. The largest webs, built by the tropical, golden orb-weavers (*Nephila* species) may be a metre across and, with support threads, can span gaps from tree to tree. A more modest and typical species is

Below: Orb-weaving spider in web with prey.

Above: Spiny-backed spider. The abdomen has a tougher exterior than in most spiders.

Opposite: Female Decorated Orb-weaver on web. The zigzags appear to increase the size of the spider.

Experiments have shown that orb-weavers can remember the position of previously captured prey. If a spider is already feeding, while a second insect gets caught in the web, the spider will wrap up the second item and leave it for later. If then, the second insect is experimentally removed, the spider will remember it and search for it in the part of the web that it was in.

Spiny-backed spiders (*Gasteracantha* species), known as kite spiders in some parts of the world, have a particularly tough abdomen. The abdomen is usually wider than long, flattened and spiny in outline. The spines reach their most bizarre in the Bull's Horn spider, *Gasteracantha arcuata* of Southeast Asia. Presumably, the spines are to make it less palatable to predators. As in the golden orb-weavers, the males are much smaller than the females. Spiny-backed spiders build their orb webs across wide gaps at heights of up to six metres above the ground. In such a position they are often quite exposed and in danger of being destroyed by birds; conspicuous tufts of silk adorn the web and are probably designed to deter birds from flying into it. The distribution of spiny-backed spiders is mostly tropical and subtropical. The genus *Micrathena*, found in the Americas, includes more species decorated by spikes or spines.

the European Garden Spider (*Araneus diadematus*) familiar to many people, especially during late summer. It has a white cross on its back and is sometimes called the Cross Spider, or Kreuzspinne. It is also found in North America.

The garden spider we tend to see is usually a stout female sitting at the centre of her web, though many prefer to be out of sight in a hideout nearby. Females have large abdomens to accommodate their eggs and silk glands. They are usually much more conspicuous in webs than males. The smaller, slimmer males appear to have a tough life and face a hazardous courtship. In the autumn, the mated females produce a yellow cocoon containing about 500 eggs and then die. Sometimes the cocoon can be found during the winter, attached to some garden woodwork. The eggs hatch, without their mother, in the spring. The young spiderlings, which stay together for a week, can be recognized by their yellow abdomen with a dark patch.

Decorated orb webs and the puzzle of the stabilimentum

Some orb webs are distinguished by their decorative markings: zigzags, spirals and bands known as stabilimenta. These flourishes are often highly visible, and are made only by spiders that sit at the centre (hub) of their web. They look like a kind of signature by the author of the web, particularly as there is considerable variation among individuals. To the human eye, they make the spider more conspicuous, and this seems strange. Various explanations for the stabilimenta have been put forward, but none is convincing. The idea that they stabilize the web is highly dubious. Perhaps they are for predator

Right: *St Andrew's Cross spider (*Argiope keyserlingi*).*

distraction, or warnings to large insects or birds not to blunder into the web. Perhaps they are connected to the spider's pumping movements when shaking the web, which often follows disturbance. The colourful St Andrew's Cross Spider, *Argiope keyserlingi* of Australia, is an example of a species that sits in the middle of the web with its eight legs arranged in a cross of four. The effect of the stabilimentum, in 'X' formation beyond the legs, is to make the spider appear larger.

Another explanation is that the zigzags reflect ultraviolet light, as flowers do, and so may be attractive to insects. In some spiders (*Cyclosa* species) the stabilimentum incorporates the jumbled remains of former prey items plus egg sacs. In this case, the assemblage appears to hide the spider, so it does seem that there may more than one explanation for stabilimenta.

Dome webs

The bulky, three-dimensional dome webs spun by members of the genus *Cyrtophora*, for example *Cyrtophora moluccensis* of Southeast Asia, illustrate a number of interesting features. Conceivably, this web is an intermediate stage between the sheet of the Linyphiidae and orb of the Araneidae. It has a fine, horizontal sheet which radiates like an orb and resembles a trampoline, with a 'knock-down maze' of scaffolding above and below. Unusually for members of the Araneidae, the web lacks stickiness. It is a robust, semi-permanent construction that is often host to numbers of kleptoparasites (see page 56). It holds prey less effectively than a sticky orb web of a similar size, it is also more conspicuous and more easily avoided by insects, but it requires less maintenance. Its chief virtue may be that it is still in place after a heavy tropical downpour, and is able to catch the moths and other insects which fly immediately after rain.

Golden orb-weavers and long-jawed orb-weavers

The golden orb-weavers of tropical and sub-tropical regions, such as *Nephila maculata* of south-east Asia, make the biggest and strongest webs – up to 2m across. The females are large, orange and brown spiders, often with feathery tufts on the legs. They are particularly disliked by unwary hikers because the webs are often built across little-used trails. However, apart from natural squeamishness when walking into a web, there is little real danger from an encounter with a golden orb-weaver. If one looks close-ly at a web, it may be possible to discern males waiting their chance to mate with the

female. Any other small spiders are likely to be unrelated and unwelcome intruders. *Nephila* is notable for building webs high up on the wires of utility lines. In such positions, under the tropical sun, these spiders are highly exposed and will orientate their bodies to minimize the heat load.

Apart from their size and golden sheen, these finely-meshed, vertical webs can be recognized because the lower half is much larger than the upper. The hub, where the spider waits, is located near the top of the web. The sticky capture thread occupies only the part of the web below the hub. It is laid by the spider in pendulum fashion, as opposed to a spiral that circles round and round. The web is not regularly renewed as in the case of most araneids. Like the web of *Cyrtophora*, it suggests an intermediate stage of evolution between sheet webs and

Left: *Long-jawed orb-weaver on web.*

"THE NATIVES OF NEW GUINEA HAVE A HISTORY OF USING THE WEBS OF NEPHILA TO MAKE FISHING NETS. A NET WAS MADE BY WINDING THREE OR FOUR STRONG WEBS ACROSS THE BENT END OF A LONG BAMBOO CANE. SKILFULLY USED, IT COULD LIFT FISH UP TO A POUND (0.45KG) IN WEIGHT OUT OF THE WATER AND ON TO THE BANK. IN THE WATER THE NET WAS INVISIBLE; IT WAS ALSO REUSABLE, AS THE SILK DID NOT EASILY PERISH."

typical orb webs. *Nephila* species sometimes build an adjacent 'barrier web', a structure that gives them some protection. The behaviour of *Nephila* appears to be somewhat primitive as prey is always bitten first, whereas *Argiope* always wraps first – a safer strategy.

Many long-jawed orb-weavers (Tetragnathidae) build orb webs that are horizontal rather than vertical. They can also be recognized because the web has an open centre. Frequently, they are found in wet, swampy habitats. The typical species, such as *Tetragnatha extensa* of Europe and North America, are slim bodied with a shiny metallic appearance and large jaws. Both sexes in *Tetragnatha* have projections on their jaws which interlock, to restrain each other during face-to-face mating.

Squatters and thieves

The big, semi-permanent webs of *Nephila*, like the dome webs of *Cyrtophora*, have the disadvantage of being liable to attract unwanted guests. Some spiders, in particular the silvery *Argyrodes* species, have given up making their own webs and have turned to invading the webs of others. The kinds of spider that prefer to take advantage in this way, rather than build their own webs, are known as 'kleptoparasites' [from the Greek *kleptes*, a thief]. In a large web, it can be difficult for the rightful owner to patrol, and keep out tiny intruders and pilferers that move surreptitiously among the threads.

Argyrodes species occur mostly in the tropics and subtropics. They are so small that they may be only one-hundredth the weight of the host spider. *Argyrodes elevatus* for example, lives in the large golden orb web of the American *Nephila clavipes,* feeding on small neglected insects or sharing the host's prey. The cheeky *Argyrodes* is also adept at stealing sizeable prey. She cuts filaments between the insect and the host spider, who is likely to remain unaware. The insect may be five times the weight of *Argyrodes*, so she secures a line to it from above, snips the last filaments and swings it away using the derrick principle! *Argyrodes* also cuts out sections of web and rolls them up to eat. The silk is nutritious and likely to be garnished with adhering pollen grains and spores.

When 30 to 40 of these little spiders occur in a single web, the host may suffer considerable losses. Often *Nephila* becomes aware of their activities and may attempt to chase them out. However, it has been observed that their numbers actually increase in the webs of female *Nephila plumipes* when courting males are present; it seems that *Argyrodes* takes advantage of the distracted female. Things can get so bad that, if the web becomes heavily infested, the host is likely to depart and build a fresh web elsewhere. Thus the kleptoparasites cause harm in a number of ways: by their steady removal of food and silk, and by forcing the host to make a hazardous departure.

Unusual forms of orb web

Besides the normal orb webs, there are some unusual forms such as the 'ladder' web, which is built by the tropical American *Scoloderus*, for example *Scoloderus cordatus*. The ladder web takes as long as three hours to build and may be up to 1.2 metres tall. It consists of an orb at the bottom with a 15cm-wide, rectangular extension above, which resembles a ladder. This astonishing web is more effective against moths than regular orb webs. Moths, because of their loose scales, may not stick to a normal orb web, but after a collision with a ladder web they tumble down and get stuck at the bottom, having lost many scales. The web is dismantled at dawn and rebuilt the next evening.

Some orb webs are noted for their particular use of space. The ornate orb-weaver,

Herennia ornatissima of Southeast Asia, builds its orb web flat against a tree trunk or rock, within a few millimetres of the surface. In this position the web is virtually invisible. Insects such as dragonflies come to land on the tree trunk without seeing the web. Many smaller orb-weavers, such as *Araniella cucurbitina* of Europe, manage to build their web within the curl of a single leaf. There is even one spider (the South American *Wendilgarda*) which lowers its web onto the surface of streams to trawl for aquatic insects.

WEBS MADE OF CRIBELLATE SILK

The webs of cribellate spiders are particularly interesting because they function without any sticky substance. The main families of cribellate weavers include the Amaurobiidae, Deinopidae and Uloboridae, which occur in

Above: Orb-weaver with woolly egg sac.

most parts of the world, and the Dictynidae, Eresidae and Filistatidae, which are found mainly in warmer regions. All produce an ultra-fine, dry, woolly silk. The silk is spun with a 'back-combing' movement, by a comb on a rear leg from a field of fine openings (called the cribellum) in front of the spinnerets. Their webs are mostly sheet-like but the uloborids spin orb webs and the dinopids make expandable nets.

Lace web-weavers

Opposite: Triangle Web Spider holds the web as a spring-trap.

Below: Net-casting spider holding it's expandable web, or net, with its long legs.

Many lace webs, for example that of the European *Amaurobius similis*, which is common on tree trunks, have a characteristic fuzzy texture to the silk and may look bluish when fresh. The silk is remarkably good at snagging the legs of insects. For this reason, *Amaurobius* and other lace web-weavers are largely immune to the attentions of the spider-hunting wasps which, unable to negotiate the fuzzy silk, are liable to get caught and eaten by the spider.

Ogre-eyed spiders

The net-casting, gladiator, ogre-eyed or ogre-faced spiders (*Deinopis* species), which occur in tropical and subtropical regions, have two enormous eyes and a remarkable method of catching prey. At rest, during the day, they stretch out their long, slim bodies along a twig, making them hard to see. After sunset, they start to build a framework of dry silk enclosing zigzag bands of fuzzy silk. The finished net, the size of a postage-stamp, is extremely elastic. Holding the net between its legs, the spider hangs down from some threads, keeping a safe distance from ground predators. When an insect comes within range, *Deinopis* drops down, stretches the net over it, and springs back.

Deinopis owes its spectacular appearance to its eyes. The two principal eyes are probably the largest simple eyes of any land invertebrate; they may be as much as 1.4mm in diameter. The eye, taking into account its relatively short focal length, has an f-number of about 0.6 (greater than most camera lenses), thus its light-gathering power is exceptional. The enormous eyes are so sensitive that the spider can hunt in near darkness. The retinal receptors are large and capable of absorbing 2,000 times more light, per receptor, than a jumping spider, most of which are active during the day. In the darkness of the forest at night, a net-caster guided by vision, like *Deinopis*, needs all the light it can gather.

Reduced webs

Having reached the pinnacle of evolution, the orb web has, in

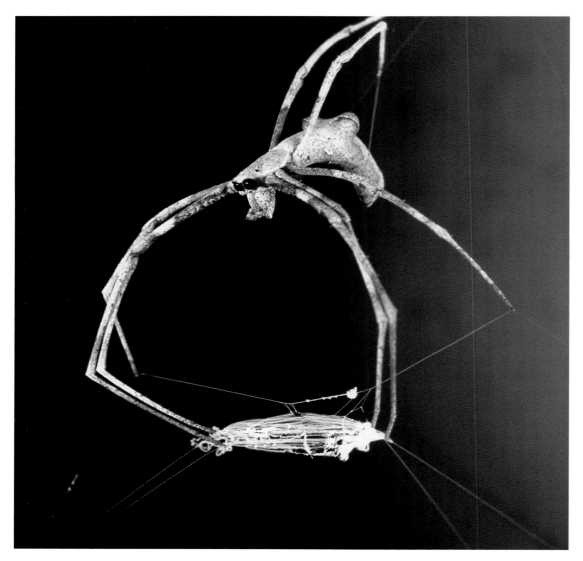

some species, turned back towards reduction and simplification. Examples have occurred in both cribellates (Uloboridae) and non-cribellates (Araneidae and Theridiidae). Some webs are reduced to just two or three threads in an 'H' shape, or even just a single thread, with or without sticky globules. The philosophy behind the reduced webs is probably to do with economy in using silk. One might expect these reduced webs to be much less effective than typical multiple-thread webs; however, they catch many more insects *per thread* than normal webs. They are highly inconspicuous and single-threads are actually attractive, as a resting place, to various unsuspecting insects.

A first stage in the reduction of the orb is demonstrated by the web of the Missing Segment Spider, *Zygiella x-notata* of Europe, in which one section, like a piece of cake, is missing. This has been replaced by a solitary signal thread that goes from the hub to the spider's hideout. To construct this web, *Zygiella* lays the spiral thread, not round and round, but back and forth in pendulum fashion. Its web can often be seen throughout the year, in the corners of windows on the outside of buildings.

Orb webs constructed by the cribellate family Uloboridae often resemble the regular webs of the Araneidae, though their silk differs. The Triangle-web Spider, *Hyptiotes paradoxus* in Europe and *Hyptiotes cavatus* in North America, builds a considerably reduced orb. This web is a kind of spring trap, under tension, consisting of three sections which form a triangle converging on a signal thread. The spider itself forms a living bridge between the signal thread, held in its front legs, and a thread going from the spider's spinnerets to an anchor point. The triangle web is one web that an insect will not disentangle itself from, because the spider collapses it around the prey with a release of silk from its spinnerets. A new web is built after each catch.

The 'H' web of *Episinus maculipes* (Theridiidae), from Europe, is a greatly reduced web, aimed at trapping insects moving along the narrow branch of a tree or shrub. The spider itself is part of the structure. Suspended by a thread from above, the spider assumes a head-down position with the first two legs holding the two lower threads of the 'H'. These two threads terminate at the branch below, where they are beaded with sticky globules to trap passing insects.

A single thread is used by the bolas spiders, so called because they resemble a South American gaucho throwing a bolas to arrest a steer. Held by a front leg, the spider dangles a line with sticky globules at the end. At night, the bolas is whirled at passing moths and when one gets stuck, the spider pulls it up, fluttering furiously. It is an amazing fact, but this spider catches only male moths. It does so by synthesizing and releasing into the air chemical compounds that mimic the pheromones emitted by the female moths to attract males from a distance. Moths, such as male Armyworm or Cutworm moths, always approach the spiders from downwind and are caught only at certain times of the night – the periods of sexual activity of the moths. During the day the spider rests on a twig and may resemble a bud, or even a bird's dropping.

Mastophora bisaccata, the American Bolas Spider and its relatives, *Celaenia distincta* and *Ordgarius magnificus* in Australia, and *Cladomelea akermani* in Africa, are all plump-bodied, often strangely ornamented spiders, which are rarely seen since they are active only at night. Some species do not even make a bolas – they simply hang from a single thread. Moths are captured when they fly straight into the spider's outstretched legs. The ultimate reduction of webs is seen in the Australian *Arkys* species. This spider builds no web at all but just sits on a silk pad, on a leaf, and seizes prey with its heavily spined front legs. It has come to resemble the crab spiders.

CHAPTER 4
TARANTULAS & TRAPDOOR SPIDERS

The largest spiders are the tarantulas, or bird-eaters. Tarantulas may seem lumbering, but the way they move is actually very graceful. They are mostly non-aggressive and rarely do they manage to catch birds. Often their diet consists simply of beetles and ants. However, the largest species are powerful enough to feed on small vertebrates such as frogs and mice. The tarantulas, and the various families of trapdoor spiders, are collectively known as mygalomorphs. They are renowned for living for 20 years or more.

BEFORE 1705, EUROPEANS KNEW NOTHING OF TARANTULAS AS BIG AS YOUR HAND. IN THAT YEAR, THE EXISTENCE OF THESE LARGE TROPICAL SPIDERS WAS DEMONSTRATED BY THE REMARKABLE MADAME MARIA SIBYLLA MERIAN, IN HER SUPERB WORK ON THE INSECTS OF SURINAM. HER ICONIC PAINTING OF A TARANTULA ATTACKING A HUMMINGBIRD BECAME THE ORIGIN OF THE SOMEWHAT FANCIFUL NAME 'BIRD-EATING SPIDER'. MERIAN WAS PRAISED IN THE EIGHTEENTH CENTURY BUT LATER, DURING THE 'AGE OF REASON', SHE WAS MOCKED AND BIRD-EATING SPIDERS WERE DISMISSED BY SCIENTISTS AS BEING NOTHING MORE THAN THE FANTASIES OF FEARFUL MINDS.

Tarantulas

Previous: *Mexican Red-kneed Tarantula.*

Below: *Metallic Pinktoe Tarantula* Avicularia metallica *seen through leaf.*

Called tarantulas in the Americas, bird-eating spiders, mygales or Vogelspinnen in Europe, baboon spiders in Africa, earth tigers in Southeast Asia, mata-caballos (horse-killers) or araños peludas (hairy spiders) in Latin America, these are the large, well-built members of the family Theraphosidae. They have heavy jaws projecting forwards and they rival in size the biggest land invertebrates, such as Emperor Scorpions, Goliath Beetles and the largest dragonflies. The largest species of all, the Goliath Tarantula (*Theraphosa blondi*) of northern South America, has a legspan in the male of up to 27cm, although its weight is no more than 85g. There are

approximately 750 species of tarantula in the world, the great majority of which are found in warm regions. In Europe there are no theraphosids (the nearest species are in Turkey and Cyprus). Incidentally, the name tarantula was originally usurped from the Italian wolf spider. Centuries ago, the name was adopted in the Americas when the European colonists met the local theraphosids.

Despite their formidable appearance, most species of tarantulas will attack a human only if provoked. Some species, however, can be irritable and aggressive, and quite capable of giving a painful bite, though

the toxicity is low and it is rarely fatal to humans. New World tarantulas tend to be relatively docile, though they are liable to rub a cloud of irritant hairs from their abdomen in self defence. By contrast, species from Africa, South Asia and Australia, for example the Usambara Baboon Spider (*Pterinochilus murinus*), the Sri Lankan Ornamental Tarantula (*Poecilotheria ornata*), and the Whistling or Barking Spider (*Selenocosmia stirlingi*), are usually more pugnacious, though they do not rub hairs in defence. If disturbed, tarantulas adopt a threat posture by raising the front half of the body, with legs

Above: Usambara Baboon Spider.

Above: *Tarantula being attacked by Giant Desert Centipede.*

Opposite: *Pink-toed Tarantula on Heliconia flower.*

held high in the air, and jaws ready for action. If necessary, they strike downwards, driving in the large parallel fangs. With such power, a highly poisonous venom is not needed.

Most tarantulas are coloured black or brown, but some are more colourful, for example the Mexican Red-kneed Tarantula (*Brachypelma smithi*), with patches of orange or red on its legs, and the Steely-blue Tarantula (*Pamphobetus antinous*) from Bolivia and Peru, with its purple iridescence. All have a thick covering of hairs, making them highly sensitive to vibrations. They are mostly active at night; their eyesight is poor and the tiny eyes can distinguish little more than levels of light. Tree-dwelling or arboreal species have dense brushes of hairs on the ends of their legs to give adhesion on the smoothest of leaves.

Tarantulas have many enemies. Mammals such as mongooses dig them from their burrows, and the young are preyed upon by birds, reptiles, amphibians and other tarantulas. They are also the unfortunate target of the Tarantula Hawk Wasp, which stings and paralyses the spider, then drags it to a burrow which is sealed after an egg is laid on the victim. Tarantulas are most vulnerable during moulting, when they are fragile and cannot move. At that time it is possible for them to be killed by smaller insects.

Forest tarantulas

Tarantulas have their headquarters in hot, steamy rainforests. Their nests and burrows may be built among the spreading roots and buttresses of trees, in bromeliads, behind banana leaves, in silk-lined spaces among leaf litter, and also in or on human dwellings. The types of tarantulas that dwell in burrows usually emerge at night and sit close to the entrance. The burrows are invariably lined with silk but do not have trapdoors. Some species build large sheets of silk webbing extending out from their nest.

Probably the most common and familiar species of all, but not the largest, is the Pink-toed Tarantula (*Avicularia avicularia*). Its

Left: *Tarantula in Arizona desert. An adult male* Aphonopelma *moving in search of a mate.*

colour is dark grey but the very ends of its legs are pink. By the standards of tarantulas, this species has a wide distribution throughout the Amazon basin and on many islands in the Caribbean. It is a good climber and will make its nest in the folded leaves of banana trees and also on the palm thatch roofs of human dwellings. A favourite spot for the nest is in the hollow centre of a pineapple plant. Besides insects, the Pink-toed Tarantula includes tree frogs in its diet.

The world's largest spider is the Goliath Tarantula (*Theraphosa blondi*). It inhabits the northern Amazonian forest and lives among the trees in a deep burrow. It digs its own burrow, or uses an old one abandoned by a rodent. The spider's hunting range is limited to a few metres surrounding the burrow. The Goliath Tarantula pounces on its prey, including insects and small vertebrates, and carries it back to the burrow. It is a solitary spider that rests in its burrow during the day, associating with another spider only when mating. The male has spurs or hooks on its front legs to restrain the female, but occasionally a male gets killed during mating. The female

lays about 50 eggs in a cocoon, which it stores in the burrow for about six weeks. After the young hatch, they remain in the nest until their first moult, and then disperse.

If threatened, the Goliath Tarantula will issue a warning. It makes a hissing noise by rubbing together bristles on its legs. It will also defend itself by flicking, with a back leg, irritant hairs from its abdomen; these barbed hairs are liable to have a serious effect if they get into the eyes or mouth. *Theraphosa* will also rear up on its hind legs in an attack position. Although not very toxic to humans, a bite can cause considerable pain, nausea and sweating.

Desert tarantulas

There are many species of tarantula that have adapted to life in the desert. In southwestern USA and Mexico, for example, the Western Desert Tarantula (*Aphonopelma chalcodes*) is able to survive in the extreme heat and dryness of the desert. It spends the daylight hours hidden in its burrow, which may be as much as 50cm deep and is frequently sited under a large stone. The

Above: *Goliath Tarantula* Theraphosa blondi. *The large bald patch on the abdomen indicates the loss of irritant hairs used in self defence.*

Above: Trapdoor spider emerging from burrow.

Western Desert Tarantula is reclusive and nocturnal, catching its prey at night and, by resting during the day, it avoids enemies such as birds and snakes. Its prey includes lizards, crickets, beetles, grasshoppers, cicadas and caterpillars. These tarantulas usually live in loose groups rather close to their neighbours but, being cannibals, are not sociable. The loose groupings occur because favourable locations are often limited and the young do not disperse far away. As the adults themselves also do not stray far from their burrow, they tend to have little interaction with their neighbours.

An individual Western Desert Tarantula may take ten years to reach maturity. Mature females tend to live in one spot and can be found in the same hole, or under the same stone, not only month after month, but year after year. Once adult, males by contrast live for only two or three months. During this time they move around a great deal and, especially during rains, can be seen crossing roads to reach new colonies. As the only mobile section of the community, it is important that males disperse some distance away during the mating season in order to prevent inbreeding.

When a male and female meet, the two rise and spread their fangs. The male secures the female's fangs behind spurs on the inside of his front legs. Often he is able to insert both palps at once into her two genital openings. Sperm can be stored for weeks or months in the female's abdomen until she is ready to lay her eggs. As they are laid, each egg is bathed in the sperm. The egg cocoons, containing up to a thousand eggs, are kept in the female's burrow and the entrance is usually silked over. The eggs hatch within seven weeks. The newborn young, which are at risk of being eaten by their mother, cut holes in the silk cover and disperse rather quickly, within a few days.

As with all spiders, the young spiderlings resemble small females. It is not until later that sexual differentiation occurs, although in *Aphonopelma* the adult sexual dimorphism and size difference is not as great as in other spiders. Both adult male and female have a legspan of approximately 10cm. Most spiderlings fail to reach sexual maturity; they may be taken by predators or else fail to find enough food to survive.

Trapdoor spiders

Trapdoor spiders (Ctenizidae and other families) are the most skilled tunnellers of all spiders. They are also the least often seen, except, that is, when roving males are found drowned after falling into a swimming pool. Trapdoor spiders occur in most warm

regions of the world. They have stout bodies with quite thick legs but are generally much smaller than tarantulas. They are usually not fussy about eating distasteful creatures. They will accept woodlice and ants, prey items that most spiders reject. However, if they do not like the insect prey, it is rapidly cast out alive and unharmed from the nest. Not all species close their burrow with a trapdoor, but those that do keep them tightly shut against predators. When hunting, some species leave the door slightly ajar with the front legs sticking out; but the smallest alarm, whether from a footfall or the flickering of a lamp, prompts the spider instantly to close the door.

Trapdoor spiders live underground in finely-constructed tunnels or burrows which they excavate using their jaws. Their jaws are furnished with a series of teeth (rastellum) to aid in digging. The burrow may be 30cm deep, situated in firm ground that does not flood, usually silk-lined and spacious enough to allow the spider to turn round. It is a haven protected from heat and rain. The entrance is usually sealed with a tight-fitting, hinged door made of earth cemented with silk and held shut with terrific force. Some spiders hold the door shut with their fangs, others use their back or front legs. Sometimes there is a second door within – one which leads to a side-tunnel. If an intruder such as a centipede gets in, the spider retreats to the side-tunnel and slams the

"A SPECIES CLAIMED TO BE LARGER THAN THE GOLIATH TARANTULA HAS BEEN REPORTED FROM THE REMOTE RAINFOREST OF SOUTH-EASTERN PERU. SO FAR, ONE SPECIMEN OF THE SO-CALLED CHICKEN SPIDER (ARAÑA POLLITO), WITH A BODY LENGTH OF 95MM AND A LEG-SPAN OF 250MM, HAS BEEN SEEN. THOUGHT TO BE A SPECIES OF PAMPHOBETEUS, IT IS SAID THAT UP TO 50 RELATED SPECIMENS SHARE A SINGLE BURROW AND THAT THEY COOPERATE IN PREY CAPTURE TO OVERCOME LARGER ANIMALS."

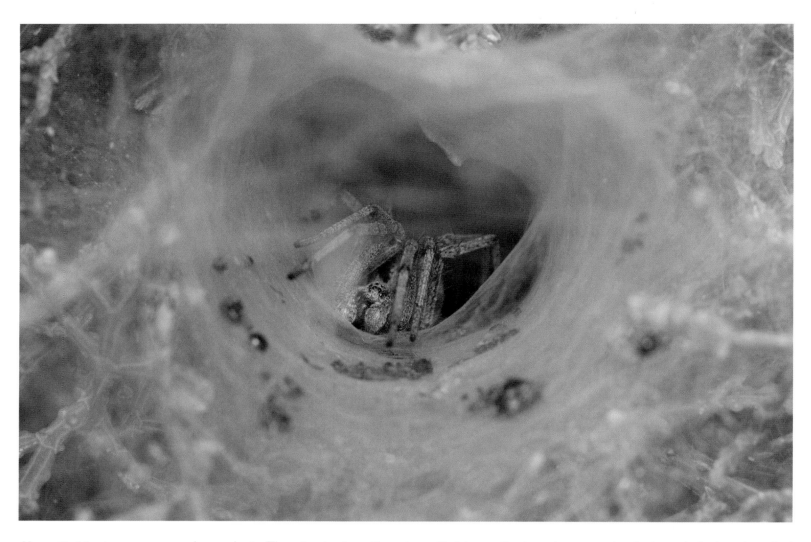

Above: *Spider in funnel-web. This type of web is built by certain araneomorphs and mygalomorphs.*

door shut. The Australian Trapdoor Spider (*Dekana diversicolor*) builds in an escape route – a second hole on the surface loosely capped with sticks and stones through which it can easily push if threatened.

The American Trapdoor Spider (*Bothriocyrtum californicum*) builds a thick, heavy door, made up of alternating layers of soil and silk, which fits in the opening like a cork in a bottle. When disturbed, the spider holds the door down tenaciously with its claws and fangs, bracing its legs against the side of the burrow. The door is difficult to open even with the aid of a knife. Researchers have measured the force using a spring-scale and found that the spider could resist a pull of as much as 0.36kg. As the spider weighs about 3 grams, the force is thus 120 times the spider's own weight. For a 70kg man the equivalent would be over 8 tonnes. However, the spider can keep this up for only a short time. So to make doubly sure, during the hottest time of the year when parasitic wasps are most active,

the trapdoor may be fastened shut and sealed with extra silk.

Sometimes the entrance is surrounded by prey detectors. These are normally silk lines but the Australian Wishbone Spider (*Aganippe raphiduca*) economizes on silk by arranging twigs which fan out from the burrow entrance. These twigs both enlarge the hunting area and assist as 'feeling lines'. Minute vibrations created by the prey's feet, as it walks overhead and crosses the twig lines, penetrate the soil clearly enough to be picked up by the spider, which is poised ready behind its door. It lifts the door, rushes out, grabs the prey and instantly returns to its burrow.

Funnel-web spiders

Funnel-web spiders (Dipluridae and other families) are mostly tropical and subtropical in their distribution. In common with the tarantulas, they have downward-stabbing fangs and two pairs of book lungs. They are

typically brown to black in colour but are not as large as the tarantulas. Also, lacking a rastellum, they are not able to dig into hard soil as do the trapdoor spiders. Funnel-web spiders are distinguished by their long, flexible spinnerets which enable them to construct a wide tubular retreat (funnel) made of silk, opening out to a large sheet web up to one metre across. They are fast-moving spiders, with the best eyesight of all mygalomorphs. They rush out from their retreat to catch prey struggling on the sheet.

Because these funnel webs, with their abundant silk, are quite conspicuous, they are much easier to locate than the nests of trapdoor spiders. In tropical regions, they can be common in road cuttings and on banks of earth. However, the family includes some notorious, poisonous species, such as the Sydney Funnel-web Spider (*Atrax robustus*). If a person disturbs the web of this spider, the pugnacious male will defend itself with a bite. The male can be deadly as the

concentration of the toxin (atraxotoxin) in its venom is six times higher than in the female (see page 117). The highly venomous atraxotoxin can cause respiratory paralysis and possibly lead to death. The Maternal Funnel-weaver (*Ischnothele caudata*) from South America is unusual among funnel-web spiders in that the female cares for the young. And when they are older, the juveniles cooperate in catching prey.

Purse-web spiders

Purse-web spiders (*Atypus* species) are among the few mygalomorphs to inhabit the temperate regions of Europe and North America. They construct one of the strangest kinds of web. The purse-web is a silken tube, looking like the finger of a glove (15–30cm long), which emerges from the soil as the extension of a silk-lined burrow, extending across the ground or up the trunk of a tree. When a fly alights on the tube it is attacked through the fabric by a massive pair

Above: *Purse-web spider (Atypus). An adult male searching outside of its purse-web for a mate.*

Above: *Giant trapdoor spider. The soft trapdoor is open allowing the spider to emerge.*

of fangs belonging to the spider which lurks inside. The victim is dragged in through cuts in the silk and later the break is mended. Normally, only the male *Atypus* leaves its burrow – when searching for a mate. Probably because they live permanently in a tube, for as long as seven years, they are noticeably less hairy than other mygalomorphs.

Giant trapdoor spiders

Giant trapdoor spiders, or liphistiomorphs, are the most primitive of all spiders and are often described as living fossils. They closely resemble the actual fossils from the Carboniferous period (370–280 million years ago). Liphistiomorphs are the only living spiders that still show the primitive segmentation of the abdomen, as in scorpions. However, they may be primitive but their appearance is most impressive, particularly

with their brick-red colour and partly black legs. They are generally larger than the typical trapdoor spiders but not as large as the tarantulas.

Fifty to sixty species of liphistiomorphs exist today and their distribution is restricted to Southeast Asia from Myanmar and Sumatra to China and Japan. In forests, they can be found on the bark of trees or on sloping banks of earth. They live in burrows or, in the case of cave-dwellers, in retreats fastened to the sides of caves. They are always difficult to find and probably most easily spotted on roadside banks. The silk-lined burrow has about six strands of silk radiating out and is finished with a softly woven, camouflaged door. The spider, when hungry, has each of its feet in contact with the silk threads. Any vibration, especially moving from one strand to another, means the arrival of prey. The spider's response is incredibly quick, as it dashes out to grab the prey and return to its burrow.

Tarantula hairs causing urticaria

Tarantulas kept in captivity can be hazardous, especially to children. This is not due to their venom but rather because of their hairy bodies. If some tarantulas feel threatened, they flick off clouds of barbed hairs with their legs. Small animals that inhale the hairs may be killed. And if the barbed hairs become embedded in a person's eye they can cause severe damage. This form of urticaria happens most often to those who look after tarantulas as pets at home. The worst thing to do is to rub the eyes after handling the spiders; indeed, it is essential to wear gloves and goggles, especially with South American species such as the Chile Rose Tarantula (*Grammostola roseum*). Eye surgeons may be able to remove some hairs from an eye but many are likely to remain embedded. The unfortunate patients are likely to need long-term medication.

Below: Tarantula shedding urticating hairs in defence.

CHAPTER 5
THE SILK FACTORY

Spiders' webs are essentially nets to catch insects. And spiders' silk is extraordinary. We humans have not produced anything as strong, light and elastic. Admiration for the fine webs of spiders was first expressed in a Greek legend when Arachne, a weaver of fine fabrics and tapestry, had the temerity to challenge the Goddess Athene to a spinning contest. Arachne won, but she was condemned to become a spider and to spin and weave for the rest of her life. But what an expert spinner, better than any other creature!

SPIDERS ARE NOT THE ONLY CREATURES TO MANUFACTURE SILK; A NUMBER OF INSECTS ALSO PRODUCE IT, MOST NOTABLE AMONG THESE BEING THE CATERPILLAR OF THE SILKWORM MOTH. HOWEVER, SPIDER SILK IS TWICE AS STRONG AS SILKWORM SILK AND, WHILE ANY ONE INSECT PRODUCES ONLY A SINGLE KIND OF SILK DURING ONE STAGE IN ITS LIFE, SOME SPECIES OF SPIDER CAN PRODUCE AT LEAST SIX DIFFERENT KINDS.

Previous: *Sheet webs in field.*

Below: *Golden orb spider (*Nephila *sp.).*

Spiders spend virtually their whole life in contact with silk threads of their own manufacture. Those that spin webs are entirely dependent on them to catch food. And a young spider is able to build its web, an amazing feat of natural engineering, without a single lesson!

Besides the construction of webs, spiders have many other uses for silk. They use it to make nests, attach trap-doors, furnish burrows, and construct egg cocoons. Some species use it for moulting or courting platforms, as support for the egg cocoons, as camouflage, or even as shields against the sun. Many web-weavers use a 'swathing band' to wrap up, or 'mummify', their prey. In some species, the female may wrap up the male after mating. In the case of certain crab spiders, for example the European *Xysticus cristatus*, it is the male that ties the female with silk prior to mating. And before copulation, all male spiders make a special sperm web onto which they deposit drops of sperm in order to charge the palpal organs. And most spiders trail a dragline behind them for safety; a spider dropping down on a dragline is a familiar sight. Also, small spiders of many kinds use long strands of silk to disperse by air (the technique known as ballooning, see page 19).

The two-dimensional, vertical orb web, resembling a wheel with spokes, is perhaps the best recognized of the various types of web. Common in gardens and elsewhere, fresh orb webs look immaculate

when glinting in the morning sunshine. At other times, they may be hung with dew, or encrusted with frost. Seeing a web stretched between distant anchor points, we may be struck with wonder in thinking about how the spider has managed to place it so cleverly. Considering all its complicated connections, angles and tensions, the orb web is one of the most fascinating objects that nature displays. This type of web is constructed by a large number of spiders, but each species does it slightly differently. Quite often, the species can be identified from particular details in the web's construction.

The variety of silks is key to spiders' success

Spider silk may seem weak to us, but if spiders were scaled up to the same size as humans, it is said that they would be able to spin a web strong enough to catch a helicopter. It is even claimed that a cable as thick as a thumb, woven from spider silk, could bear the weight of a jumbo jet. Spider silk is very strong, but it is also extremely fine. A typical strand of orb-weaver silk has a diameter of about 0.003mm, approximately one-tenth the diameter of silkworm silk. Of the various kinds of spider silk, the dragline of the relatively heavy golden orb spider (*Nephila* species) is believed to be the strongest natural fibre known. A dragline would need to be 80km in length before breaking under its own weight.

Spider silk is very durable and is insoluble in organic solvents. Strong acids are needed to break it down. And old abandoned cobwebs are only slowly attacked by fungi and bacteria, probably because of their acidity. However, spiders' silk (unlike nylon and silkworm silk) is affected by rain and absorbs water, causing it to swell; it returns to normal after drying. Wet threads can stretch up to three times their length before breaking, but dry threads are more brittle and may not stretch more than a third of their length before breaking.

The silk glands fill much of the rear end of the abdomen. They connect to the spinnerets, each resembling a showerhead with numbers of tiny nozzles (spigots). Silk emerges as extremely fine strands which combine into a single thread. A sudden increase in blood pressure forces liquid silk out of the nozzles. The spider pulls it with a leg, causing the chains of silk molecules to bond together and solidify. The conversion from a soluble, liquid form into an insoluble, solid form is due to a change in the orientation of the molecules. At the same time, the molecular weight of the soluble protein is increased considerably in the solidified silk. Silk becomes solid not by drying in air but by

Below: Orb-weaver's spinnerets.

Right: *Orb-web with threads bowed down by the weight of the dew.*

mechanical pulling. Drawing threads out on the breeze, or dropping down on a dragline, has the same effect. The European and Asian Water Spider (*Argyroneta aquatica*) builds its silken diving bell under water, thus proving that silk is not air-dried.

An orb-weaver has an impressive ability to change the properties of the silk it produces for different tasks. The nozzles are muscular and can change the diameter, strength and elasticity of the thread. Furthermore, up to seven different types of gland within the spider's abdomen secrete silk of varied properties for specific purposes. The main types of silk gland are: *ampullate* for the dragline and frame threads of webs; *piriform* for the bonds between separate threads (attachment disks); *aciniform* for the swathing bands to wrap prey, the male's sperm web, and the woolly silk used in the stabilimenta that adorn some webs; and *tubuliform* (absent in the male), to produce silk for the egg cocoon. In orb-weavers there are additional glands: *flagelliform* for the sticky spiral; and *aggregate* for the glue of the sticky globules (also found in tangle-weavers). As much as 700 metres of web silk may be drawn continuously from a single well-fed golden orb spider (*Nephila*).

Building an orb web

Many of us have watched with fascination as a spider builds its web, but rarely has anyone seen the whole process from the very start. The fact is you need more than luck and perseverance to witness construction from the beginning, because it is quite unpredictable. Even those who are present when the first threads are laid have difficulty in describing what they have seen, as the spider acts in a seemingly random fashion with no fixed pattern. The early stages are the least understood and the spider's movements are often interrupted by long pauses. It is as though the spider is deep in thought!

The favourite time to begin construction is an hour or so before dawn. The spider moves

Above: *Orb web spiders. The male (bottom left) is biding his time before approaching the female (top right).*

from her resting place and climbs to a prominent position. She may have used it many times before. If so, she is likely to build her web using some remaining frame and support threads from the old one. But if she is starting to build completely from scratch, then the first task is to create a line, like a tightrope, that bridges a gap. The spider initiates the line by dropping down on a dragline and climbing rapidly back up to produce a loop which is light enough to float on the air. If the air is moving, more thread is then drawn out from the spinnerets. The spider waits, as if fishing, until, with luck, the line catches on an anchor point, for example a twig or fence, some distance away. It may be a metre or more away, and the line may span a path or even a stream. Feeling the line attached to something, the spider tightens it and walks across, trailing behind her a thicker line (*bridge thread*). This becomes the basis for her construction work.

The next stage is variable and the spider behaves in a flexible manner, depending on the immediate environment. Unless she is over water, she is likely to drop down on a line from the middle of the bridge thread until, swinging perhaps in the breeze, she contacts an anchorage below and fixes the line there. Pulled tight, the resulting 'Y' or 'X' structure marks the origin of the centre (hub) and the first spokes (radii) of the future orb web.

After adding any necessary frame and support threads, the next stage establishes the radii which radiate out from the hub to the edge of the orb outlined by the frame threads. Depending on the species of spider, a web has from 10 to 80 radii and the angles between them are fairly consistent. In the case of the European Garden Spider (*Araneus diadematus*), the approximately 30 radii are placed about 12 degrees apart. Each radius is produced twice, the first temporarily in the outward direction; then cut and entirely replaced by the permanent thread laid on return to the hub. While the spider is busy placing the radii, a complex of

threads develops at the hub, which is where she may sit when the web is complete. The hub is reinforced and often ringed by three or four circular threads.

The next task is to lay a temporary spiral thread. It starts a little way out from the hub and goes round about 36 times, crossing the radii and out to the margin of the orb. It ties the radii together and serves as a non-sticky guide-line for the permanent sticky spiral, or capture thread, which is laid on return to the hub (simultaneously cutting out the temporary spiral). Laying the sticky spiral is the longest operation of all and the stage that we are most likely to observe. But understanding the precise technique needs careful observation. What actually happens is that, as the spider moves around the web, it reaches for the next radius with a front leg, while a back leg pulls a length of sticky thread from the spinnerets, dabs it against the radius to make a joint, and then gives a tug so that the thread breaks up into a line of sticky beads.

The sticky spiral usually has several U-turns, mainly along the outer edge, where the spider has reversed direction by 180 degrees. Going towards the centre, the spiral does not actually reach the hub but leaves a free zone, crossed only by radii, which allows the spider, when necessary, to dodge from one side of the web to the other. The final stage of construction is to further reinforce the hub, although in horizontal orb webs the hub is often removed to perform the same function as the free zone. Incidentally, vertical orb webs are usually not symmetrical, as the hub tends to be placed above the centre (but see page 57). This is because the spider can move faster downhill than up.

The whole process of spinning an orb web may take an industrious spider no more than an hour or so. However, they can be very easily disturbed and, if so, will simply stop building. The typical web of an *Araneus diadematus* consists of approximately 20 metres of silk (6m non-sticky, 14m sticky) interconnected by approximately 1,500

Above: *Completed orb web. The barbed wire acts as a convenient frame. The orb's hub is an open one.*

Right: *European wasp spider* (Argiope bruennichi) *out of its web.*

joints. During the web-building process, the spider continually makes adjustments, spreading the stresses and tensions evenly across the threads.

When the web, with all its structure and support threads, sticky threads and last touches are finally completed, the spider settles on the hub. Many, however, retire to a nearby hideout, protected from predators, and remain in touch through a 'signal thread' held by a front leg. The webs are marvellous constructions but many are demolished by their builders each night, because of daily wear and tear and loss of stickiness, and are rebuilt in the hours before dawn to reduce exposure to predators. But in tropical regions, many orb-weavers renew their webs up to five times within 24 hours, after each downpour. The old web is rolled up and ingested, together with tiny insects and adhering pollen grains. Within two hours, the silk protein is re-absorbed and recycled, ready for use as new silk. Often the new web re-uses the anchor and some frame threads of the old web, but the radii and sticky spiral are always renewed.

An orb-weaver may build as many as 100 webs in its lifetime, each one weighing less than a milligram. Webs are extremely light; rolled into ball, one forms a mass about the size of a grain of rice. And yet it has been estimated that, in an acre of heathland, all the spiders living there own 12,000 miles of silk (the distance between Europe and Australia).

How do orb webs capture prey?

In favourable weather conditions, as many as 250 insects a day can be caught in a single orb web if it is well sited relative to the prey. For example, in meadows, most insects stay close to the ground, whereas in forests, more insects fly somewhat higher up. Therefore spiders make complex calculations such as: 'how big is the open space?', 'how much silk do I have?', and 'what anchor points are available?' Indeed, they have an impressive ability to act according to the circumstances.

For example, an orb web can be built using just three anchor points, but the spider will use more if available. And if the web in a particular site fails to provide enough prey, then the spider may move on. The behaviour of spiders is all the more impressive when one considers that they cannot see objects but depend essentially on their sense of touch. It even seems that spiders can learn a spatial map of the surroundings so that subsequent webs, in the same place, need less exploration.

Above: *Florida Garden spider (*Argiope argentata*) wrapping prey in web.*

85

Above: *Decoy Spider*
Cyclosa insulana
camouflaged in
decorated web.

The ability of orb webs to stop and retain plenty of insects is influenced by many factors in their design: the density of silk threads, the strength of the material, the distribution of tension among the threads, the ability to absorb momentum, and the quality of the adhesive. For example, webs with a higher density of radii can absorb greater energy and are able to catch heavier and faster-moving insects. They also collect greater numbers of smaller insects. However, a dense web is more visible during the day and is more extravagant in the use of silk. The protein used in the silk is an expensive resource, so webs must balance productivity with costs.

Let us consider some of the essentials. A web must resist the impacts of prey, detritus, wind and rain. It needs to be able to prevent the escape of flying or jumping insects for at least five to ten seconds, until the spider arrives to attack. From the moment of impact, an orb-weaver, for example the European and North American *Araneus pyramidatus*, takes about five seconds to bite and begin wrapping a insect. Webs that restrain prey for longer periods have an

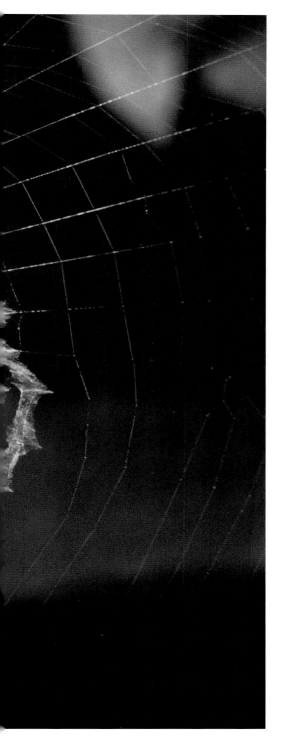

advantage, because spiders need extra time to approach large and vigorous insects. Different species of spider vary in their attack behaviour. Some bite first and then wrap the victim in silk. Others prefer to wrap first, which is useful when dealing with stinging insects such as ants. Very aggressive insects, and those that discharge noxious chemicals, are likely to be cut out of the web by the spider.

Whereas the radii must be strong, and relatively non-elastic so that the web retains its orb-shape, even when blown by the wind or

put under strain by the struggles of prey, the capture spiral must be sticky enough to hold the prey, and elastic enough not to break, when a heavy insect canons into it. Equally, it must not be too firm, like a trampoline. Otherwise an insect may well be bounced off and be lost. So the capture line must dissipate energy and return to its original length with little recoil. The specification seems almost impossible, but is achieved by its amazing microscopic structure.

The mechanism involves covering the fibres, as they emerge from the spinnerets,

Above: Orb-web with trapped dragonfly. Web and prey are beaded with water droplets. The lower part of the web is damaged but the spider shows no interest in consuming the insect.

Above: *Orb-web spider (Nephila) with sunbird caught in web.*

with a coat of liquid glue. As each length is fastened to a radius, the spider tugs it with one of her hind legs. The vibration causes the liquid coat to coagulate into a line of equally-spaced droplets. The droplets take up water from the atmosphere and their surface tension encloses coils of fibres. When an insect hits the web, these fibres are pulled out of the droplets like microscopic fishing reels. But when the impetus of the insect is slowed to a halt, the surface tension rewinds the fibres back into the droplets and the thread reverts to its original tautness.

"ORB-WEBS TEND TO BE BUILT AROUND 4 AM IN THE MORNING. FOR TWO SCIENTISTS, PETERS AND WITT, WHO STUDIED WEB-BUILDING BEHAVIOUR IN THE 1950S, THIS HOUR WAS MOST INCONVENIENT. SO THEY HAD THE IDEA OF FEEDING STIMULANTS TO THE SPIDERS IN THE HOPE OF ADVANCING THEIR BODY CLOCK. UNFORTUNATELY, THERE WAS NO CHANGE IN THE HOUR BUT THE WEBS BUILT UNDER THE INFLUENCE OF DRUGS WERE A COMPLETE MESS. AMPHETAMINES, LSD, CAFFEINE AND OTHERS WERE TESTED. EACH CAUSED CHARACTERISTIC DISTORTIONS BUT THE MOST MARKED WAS CAFFEINE."

Orb webs made by cribellate spiders

Cribellate spiders produce silk from a special organ, the cribellum. This is a small plate in front of the spinnerets which is densely covered by many micro-scopic spigots. Extremely fine threads are combed out by a brush on the hind leg to make the dry and woolly silk. It has a core fibre but instead of being coated with sticky fluid it is surrounded by a mass of very fine fibres of dry silk, only 0.00001mm thick. The resulting silk often has a bluish appearance when fresh. It lacks the beading of gluey silk, but still has a great ability to snag the legs of insects. Indeed it has been described as the 'original Velcro'. Its stickiness is partly due to electrostatic forces.

Arachnologists have argued for years over whether cribellate spiders and non-cribellates have evolved the ability to spin orb webs independently or whether they share an orb-weaving ancestor. Considering their greater diversity, the non-cribellate spiders are the more successful group. They make the typical orb webs. Nevertheless, the silk of cribellate spiders is so effective that the one family of cribellates that spins orb webs (the Uloboridae) is the only one of the 110 families of spiders that has no need for poison

glands. They rely entirely on wrapping with cribellate silk to subdue their prey.

The advantages of cribellate silk are that it weathers well and does not need frequent renewal, as do the typical webs made with gluey silk, which catch more prey but quickly lose their stickiness. Orb webs made with cribellate silk, for example by *Uloborus* species, lie in a more or less horizontal plane but otherwise are similar in design to those made with gluey silk. The uloborid's web lasts longer but requires much more effort to produce the silk. In fact the main disadvantage is that the finely combed bands of cribellate silk take a considerable time to produce, so speed is limited. However, in one respect cribellate orb-weavers are more successful, as they can be found in buildings. Unlike typical orb-weavers (non-cribellates), they will spin their webs in the hostile, dry, northern, winter *indoor* environment.

Below: Triangle-web spider holding the old web rolled up as a ball of silk.

CHAPTER 6
MATING AND BREEDING

The coupling of two predatory and often short-sighted creatures can be a hazardous affair, particularly for the smaller male. In spiders, the battle of the sexes is very intense and, during courtship, the male works incredibly hard; elaborate and sometimes lengthy courtship rituals are often essential to overcome the cannibalistic tendencies of the female. But it is the intensity of the struggle that makes this behaviour so fascinating. During mating there are a number of important forces at play. First, there is the need for mate recognition among the thousands of species worldwide. Then there is the need for the male to avoid being mistaken as prey, given that the female is usually larger. There is also the need for the hard-working male to be selected from among his competitors, as well as the necessity to stimulate the female into becoming sexually receptive rather than just hungry. Finally, there is the competition to ensure that no other male can fertilize any of the female's eggs.

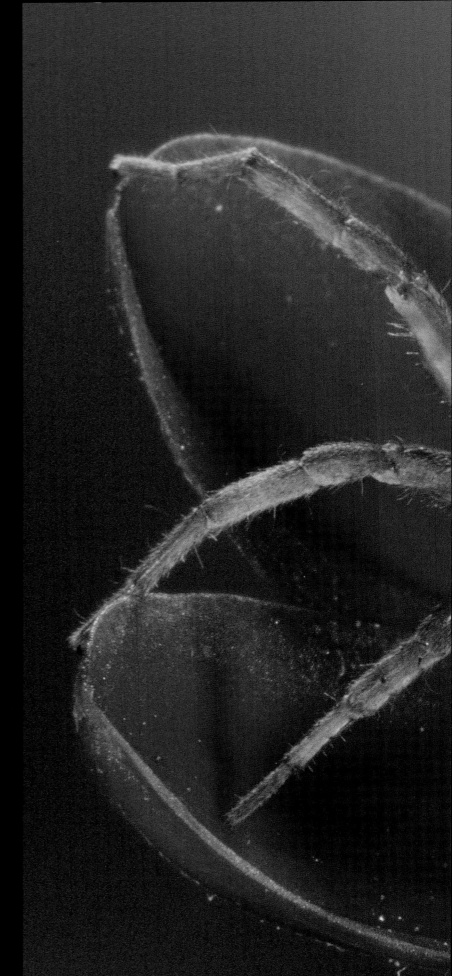

SPIDERS ALWAYS HAVE SEPARATE SEXES. MALE SPIDERS DO NOT PROVIDE PARENTAL CARE BUT THEY DO INVEST A LOT IN THE REPRODUCTIVE PROCESS. THEY ARE THE ONES WHO DO THE SEARCHING FOR A MATE, WHILE THE FEMALES STAY PUT. AND DURING COURTSHIP, MOST MALES EXPEND YET MORE ENERGY. THEY ARE OFTEN FORCED TO COMPETE WITH OTHERS, AND EVEN FIGHT, FOR ACCESS TO THE FEMALE.

Adult males can usually be distinguished from adult females by their smaller size and slimmer body but relatively longer legs. This sexual dimorphism is most obvious in many tropical web builders, such as the golden orb-weavers (*Nephila* species), in which the females are very large (up to 45mm body length) while the males are usually much smaller (4–8mm).

One technical problem shared with other early pioneers of the terrestrial environment, such as millipedes and dragonflies, is that the testes are not connected directly to copulatory equipment. To overcome the problem, male spiders use a complicated method which is unique among arthropods. It involves the transfer of sperm via a pair of accessory sex organs and a 'sperm web', which resembles a small napkin. The two

organs, carried by the palps on each side of the mouth, comprise a bulb and duct structure that functions as a pipette to take up and transfer the sperm. The male's accessory organs are themselves referred to as 'palps'.

THE MATING PROCESS

To begin the mating process, a drop of sperm is extruded from the male's genital opening on to a newly-built sperm web and then taken up by the palps. When the palps are fully charged, the male concentrates on finding a mate. In most cases, this is made possible by following pheromones, species-specific chemical signals on trails of silk, previously laid by the female. Among web-spinners, which generally have poor eyesight, the male begins an often lengthy

Previous: *Female Nursery Web Spider with egg sac on flower.*

Right: *Male Orchard Spider mating with newly moulted female.*

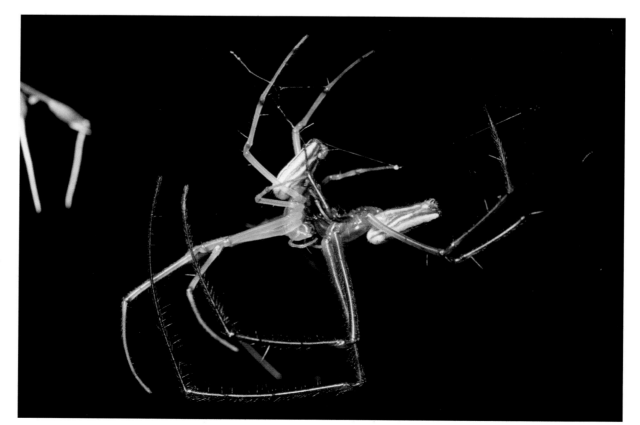

courtship ritual after contacting the female's web. In others, such as crab spiders and sac spiders, there is little courtship behaviour and mating is triggered simply by direct physical contact between the male and female. But in spiders with good eyesight, such as jumping spiders and lynx spiders, courtship starts with sight recognition and progresses to an elaborate display of visual signals. Then, if courtship is successful, the next stage is copulation. During copulation, the sperm is transferred to the female by inserting the tip of a palp into one, or both, of her pair of genital openings (*spermathecae*) and squeezing the bulb. The two palps are inserted into the two openings successively (first and second palpal insertions) or, in the case of the more primitive spiders (mygalomorphs), both together. In the relatively advanced spiders (araneomorphs), the basic bulb and duct have evolved into a highly complex organ, which combines expandable soft tissue with various hardened appendages. When the palp is inflated by hydraulic pressure these appendages project and interlock with fea-

tures on the female's genital plate (*epigyne*). They fit together (in the same species) like a 'lock and key'. Copulation lasts from a few seconds to many hours. Among the orb web-weavers, courtship is usually lengthy but copulation is brief. Within a few weeks after mating, many females are ready to begin laying eggs.

MATING STRATEGIES OF VARIOUS SPIDERS

It is interesting to note that cannibalism associated with mating is common among spiders (and scorpions) but rare among other creatures. Apart from spiders, the best known examples are a small number of insects, most notably the praying mantis and the sage bush cricket. The black widow spider (*Latrodectus* species) is famous for eating her mate but only a minority of male spiders actually die at the hands of their mates – whether it be from female predation or simply from exhaustion during mating. Nevertheless, it is a fact that males are generally smaller than females

Above: *Male black widow spider with a pair of conspicuous palps.*

Above: *Male black widow spider inserts a long, coiled duct into the female's genital opening.*

and so many species have strategies to avoid sexual cannibalism. In some species, however, the nutrition that the female obtains from eating the male (either during or after mating) gives an evolutionary advantage.

How then does a male spider get close enough to copulate while avoiding capture? Probably the simplest tactic is to wait until the female has caught an insect. The male Autumn Spider (*Metellina segmentata*) depends on this strategy, for otherwise he seems quite unable to pacify a hungry female of his own species. So he watches and waits at the edge of her web, motionless for hours or days, until she catches a relatively large prey item. Knowing that she will be satisfied for a period of time while consuming it, he then takes his chance. He enters the web as she inserts her fangs into the prey. At the same time he begins to sig-

nal his identity and proceed with courtship.

The male Nursery Web Spider (*Pisaura mirabilis*) goes one stage further. First, he catches an insect. Then he gift-wraps it in silk, something he would not normally do. Carrying it in his jaws, he goes in search of a female. On approach, he straightens his legs and tilts his body up to present the gift, safely placed between him and her jaws. She bites into it. And as she does so, he ducks under her abdomen and inserts his palps. While she is dealing with the nuptial gift, it is relatively safe for him to copulate. Essentially, the gift protects against sexual cannibalism but, as food, it can also be considered as the male's contribution to the reproductive process. Furthermore, the bigger the insect (or the greater the silk wrapping!), the longer the time needed to consume it. Thus mating is prolonged and the proportion of eggs fertilized is increased.

Right: Male approaches the larger female black widow spider. For the male, mating is a stronger instinct than self preservation.

Above: Male Nursery
Web Spider presents gift
to female.

In the case of web-spinners, it is usually too dangerous to try to tiptoe across the female's web, for every vibration will tell the owner where the male is. So, to begin courtship, the male Garden Spider (*Araneus diadematus*) attaches a silk line, or mating thread, to the female's orb web. He then spends many hours plucking the line to coax her on to it from her normal resting position. The object is to confine her to the mating thread in an attempt to make her less dangerous. The pair may spend as much as a day lunging and retreating from each other until eventually the female is too tired to further resist the attentions of her suitor. Garden Spider females are in fact quite capable of cannibalizing males *before* copulation but, of course, if she is too aggressive to suitors, the risk is that she will die without mating. Later in the breeding season things can become even more difficult for the males. They usually face a totally hostile reception from females who mated long ago and require peace and quiet to feed and lay their eggs.

Courtship involving sound and vision

In the jumping spiders (Salticidae), which have the best eyesight of all arachnids, courtship displays are usually performed face to face. The males, often brightly coloured, 'dance' in front of the females, waving their legs and palps, and assuming special poses. They are charming, pretty little spiders that show a high level of awareness. However, in spite of their prettiness, some jumping spiders are quite formidable predators, such as those of the genus *Phidippus*, in North America.

Females of the species *Phidippus rimator*, for example, are fiercely territorial. If a female is placed within the visual range of an established female, a fight to the death is most likely. Females also commonly attack and consume males before, during, or after copulation. Males are thus obliged to approach very cautiously, with an elaborate courtship display. And they retreat rapidly after copulation, usually by curling into a ball and rolling down to the vegetation below. Furthermore, a male may locate a female but will not approach her until she has captured a prey item. However, a male may be killed during copulation when the female drops her prey and turns to attack her mate. Because aggressive females will better defend their eggs and foraging space from competitors and predators, such high levels of aggression are undoubtedly of survival value.

For some spiders, sound plays an important role in courtship. It is perhaps surprising that as many as a quarter of all spider families contain species capable of producing sound. Sound is produced via three distinct methods: stridulation, percussion, and vibration. Stridulation, in the style of grasshoppers, involves rubbing together various body parts, including the abdomen, jaws, palps and legs, to produce sounds. For example, some tarantulas have a file and tooth system on their jaws and palps, which

can make an audible, defensive hiss. Percussive sounds require no specialized organs and can, for example, be produced by palps or legs drummed against wood, leaves, webs, or even water. Some, including the large tropical banana spiders (Sparassidae), can produce humming sounds with whole-body vibrations or rapid leg oscillations, which vibrate through the vegetation.

The noisiest are the wolf spiders (Lycosidae). In many species, an excited drumming of the palps, audible to the human ear, occurs as the male follows the pheromone trail of the female. To identify himself and impress the female, a male produces various buzzing sounds by means of stridulation. These sounds require ultra-sensitive recording equipment and are not audible to the human ear. Their frequency range is quite low, around 250–1000Hz, and it is the pattern of the sound waves that is most

Left: *Buzzing Spider (Anyphaena accentuata) vibrating its abdomen against leaf to attract mate.*

Left: Male jumping
spider (on left) displays
to female.

Above: Male jumping spider in display pose.

important. As an example, two species of wolf spider look almost identical but they do not cross-breed and can be distinguished by sound. The male of the North American *Schizocosa ocreata* makes a kind of purring sound lasting several seconds. The other species, *Schizocosa rovneri*, produces a quiet buzz in short bursts separated by long pauses. The female makes no sounds, having no stridulatory organs.

Some of the strangest examples

We have seen one or two examples of how cannibalism can be a serious threat to courting males. Males are also up against the fact that females are not necessarily faithful to their partner. Males therefore have a range of strategies to ensure that their sperm is successful and that no one else will have a chance to mate with the female. Some of their tactics are rather shocking and could be described in human terms as bondage, rape, chastity belts, castration, and suicide. If a male has to compete very hard to mate, and for the one chance to pass on his genes to

the next generation, then it is likely that he will overlook any risks to his life, particularly when females are relatively few. Only a handful of species go as far as suicide but, strange as it may seem, in terms of being a father a male can benefit from being cannibalized via an increased fertilization rate.

One strategy is to disable the female. For example, male big-jawed orb-weavers (*Tetragnatha* species) are not afraid of sex because the spurs on their fangs wedge open the female's jaws so that she cannot bite during their embrace. A more bizarre example is the form of bondage practised by the male European crab spider *Xysticus cristatus*. He approaches tentatively but, when close to the female, grabs one of her legs. Initially she struggles but later calms down as he moves over her body trailing silk threads, which bind her to the ground. He then lifts her abdomen, crawls under and inserts his palps. However, this bondage appears to be purely ritualistic because it is not difficult for the female to break free; it is likely that it helps to pacify her.

Cohabitation between an adult male and an immature female is a strategy used by sac spiders (Clubionidae) and others. Such males tend to spend much of their lives focused on choosing and guarding their mates. They mature early and stay near the female for a number of days before her final moult to adulthood. Mating occurs immediately after her final moult, when she remains defenceless until the new exoskeleton hardens.

In the case of the North American Fishing Spider (*Dolomedes triton*) the battle of the sexes is so uncompromising that the female eats her mate *before* mating. Thus only exceptional males, strong enough to escape, will go on to have sex, with the result that the strongest genes are passed on. Failing that, copulation may be possible only in the short time during which the female is waiting for her exoskeleton to harden.

The golden orb-web weavers (*Nephila* species), of tropical and subtropical regions, are known for their interesting sexual behaviour. Males vary in size from a twentieth to a hundredth of the size of the female and, being so small, can avoid being taken for food. They are also able to climb over the female's body without any fear. They have no need for mating threads. But females regularly copulate with more than one mate, and rival males compete fiercely among themselves during courtship. So, at the end of copulation, males often break off the tip of a palp, which then remains lodged deep inside the female's genital opening (spermatheca), as a kind of 'chastity belt', or mating plug, to prevent further mating.

Among the decorated orb-weavers (*Argiope* species), the female is usually about ten times bigger than the male. In the American Garden Spider (*Argiope aurantia*), the male usually dies while joined in copulation with the female. When using his second palp he suffers an irreversible seizure and is dead within fifteen minutes, after his mission is accomplished. Still attached, his body acts as a large mating plug. Other males may des-

perately try to pull it out but they usually fail. The female eventually eats it. In the related species *Argiope aemula* of Southeast Asia, the female gently wraps the male in silk as he copulates. He resists if this happens during the first palpal insertion, and sometimes he is able to break free, but may lose a leg or two. However, if this is his second palpal insertion, or second mating, he often seems to lack the strength and does get eaten.

Mating plugs also occur in the widow spiders (*Latrodectus* species). The ultimate prize

Below: Male golden orb spider approaches much larger female.

Right: *Male golden orb spider inserts palpal duct into female's genital opening.*

is that males who successfully copulate with both palps, losing a tip in each of the female's spermathecae, may be expected to fertilize all of the eggs. Towards this end, the male Red-back Spider of Australia (*Latrodectus hasselti*) actually commits suicide during copulation, by somersaulting into a position where his abdomen is at the jaws of the female. Thus, by being consumed, he prolongs the act of mating, helping to ensure that no other male will fertilize any of the eggs. In fact, the determination can be so strong that two males will fight to snatch each other out of the female's jaws. As we have seen, all male spiders have two palps (accessory sexual organs for the transfer of sperm) which they insert into the female's spermathecae when mating. But one of the most bizarre and unexplained cases is that of the male tangle web-weaver *Tidarren* of the Middle East which, prior to mating, amputates one of his own palps. He then constructs a mating thread to the female's web. This is usually followed by copulation, but with only a single insertion in only one of the female's spermathecae, before dying of fatigue. The female then always eats him. She may have a number of mates but males are always monogamous due to mate consumption. This strategy is puzzling because it seems to make little sense from an evolutionary point of view. The male neither avoids cannibalism nor prevents other males from mating with the female's second spermatheca.

Clearly, it is usually the males that get a rough deal in the world of amorous spiders. But it comes as some relief to find that there is one example where the male does get the upper hand. The sheet-web weaver *Linyphia triangularis* is indeed an unusual case, where the male is dominant. He tolerates no resistance from the female, who is about the same size as himself. If prey falls on her web, the male chases her away and feeds alone. He even destroys most of her web after mating, to prevent other males from being attracted to the pheromones on her silk. Furthermore, he ensures that copulation takes longer than needed to fertilize the female's eggs. Thus most female *Linyphia triangularis* are monogamous and their partners do not need to pick a fight with other males.

"THE AUTHOR RECALLS STUMBLING UPON A DRAMATIC SCENE INVOLVING ONE FEMALE AND TWO MALE DECORATED ORB-WEAVERS (ARGIOPE BRUENNICHI). THE FEMALE, HANGING SIDEWAYS AT THE CENTRE OF HER WEB, LOOKED DISTINCTLY SICK. THE TWO MALES IN CLOSE ATTENDANCE APPEARED TO BE PRYING INTO SOMEONE ELSE'S GRIEF. THEN, SUDDENLY, THE FEMALE'S EXOSKELETON CRACKED, SHE MOULTED AND BECAME ADULT. THIS WAS FOLLOWED BY A RACE AMONG THE MALES TO BE THE FIRST TO GET TO HER. IN A SHORT SPACE OF TIME THE END RESULT WAS: ONE SHED EXOSKELETON; ONE NEWLY MOULTED AND MATED, ADULT FEMALE; ONE MALE THAT HAD LOST A LEG IN THE BATTLE; AND ONE DEAD, WRAPPED-UP MALE THAT HAD MANAGED TO MATE WITH THE FEMALE."

THE NEXT GENERATION

Depending on the species, a pregnant female spider lays from one egg to a thousand or more. Using sperm stored since copulation, the eggs are fertilized just prior to laying. They adhere together and are usually enclosed in a silken sac, or cocoon. Cocoons give protection from fluctuating temperatures and humidities, and also from predators and parasites. However, they may be breached by parasitic insects that lay their eggs among those of the spider. Thus many spiders make several cocoons, containing large numbers of eggs, to ensure that enough survive.

Exceptionally, some species, such as the Spitting spiders (p.43) and the Daddy-long-legs spiders (p.47), simply tie their eggs with a few threads and hold them in their jaws. Such spiders tend to live inside buildings and are little exposed to the elements. In most other species, the egg sac is quite substantial and often bigger than the spider itself.

Basically resembling a pie, with a spongy base and a circular wall, the eggs are deposited inside in a continuous stream and fully enclosed with a sheet woven on top.

Egg sacs vary in appearance from species to species (p.17). They may be round, flattened, knobbly, or even lantern-shaped. The colour may white, blue, green or brown, and some are camouflaged with a covering of mud or bits of bark. In general, the cocoons of hunting spiders have a tough, papery exterior while those of web-builders are often enveloped by a mesh of woolly threads. Cocoons may be fixed to the vegetation (e.g. typical orb-weavers), attached to the web (e.g. funnel-web spiders), placed under bark or stones (e.g. huntsman spiders) or kept inside a burrow (e.g. trapdoor spiders). In some cases, the eggs are laid inside an 'egg nest' which doubles as a retreat where the female rests (e.g. sac spiders and jumping spiders).

Below: Wolf spider carrying young on her back.

The habit of staying with the eggs and guarding them until the spiderlings emerge is very common in female spiders. Among those with a nomadic lifestyle, the mother carries the egg sac with her. For example, Wolf spiders attach their cocoon to their spinnerets and, after hatching, allow the young to ride on their abdomen (p.28). The Nursery Web spider (p.30), carries its large egg sac under the body but later, the mother spins the nursery web, or protected space for the young, around the sac and stands guard nearby.

The female Green Lynx spider *Peucetia viridans* (p.38) is a good example of a devoted mother who stays with her egg sac. She fastens it with guy-lines to the vegetation and defends it aggressively if threatened. Ants are a serious enemy and,

to prevent their attack, she cuts all but one or two of the guy-lines so that the egg sac swings in mid-air, while she balances on top like a trapeze artist. After the summer, when the young are ready to emerge, she spins a canopy over the sac and opens it up. But the point comes when she dies and leaves the young to survive the winter on their own.

The outward appearance of a cocoon remains largely unchanged while the spiderlings develop within. The tiny creatures that hatch are initially colourless and lacking spines. Often they feed on unhatched (infertile) eggs in the batch. They stay within the cocoon and there they undergo their first moult. Days later, they emerge, having cut holes in the sac with the jaws (helped by silk-digesting saliva or the mother outside). At first, the spiderlings

Above: *Orb-weaver with woolly egg sacs attached to leaves.*

Overleaf: *Spiderlings following emergence from an egg sac.*

105

cluster together, but whether or not they receive any maternal care, depends entirely on the species (p.17)

Many spiders, instead of spending time caring for a single batch of eggs, opt to produce multiple batches over a period of time and this usually means that the eggs will be unattended. For example, the Garden Spider of Europe and North America (p.52) dies in the autumn after producing eggs – usually six months before the young emerge in the spring. On a warm spring day, the emergence of a couple of hundred young Garden Spiders can attract attention because the cluster of tiny orange bodies, with black triangles on their backs,

appears to 'explode' and scatter at the slightest disturbance.

After a week or more, and sometimes a second moult, the spiderlings begin gradually to disperse. Dispersal is necessary to avoid competition and cannibalism among the hungry siblings. Many burrow-dwellers simply walk across the ground, but most others disperse by bridging and ballooning. Bridging is a means of travelling by repeatedly climbing up through foliage and then dropping down on a silk line to cross adjacent branches, often with some breeze-assisted swinging. Ballooning (p.19) involves climbing to a prominent point and letting out silk lines to catch the breeze and lift the spider away.

Above: Raft spider with egg sac.

Opposite: Nursery Web Spider with nest of young.

CHAPTER 7
THE USE OF VENOM

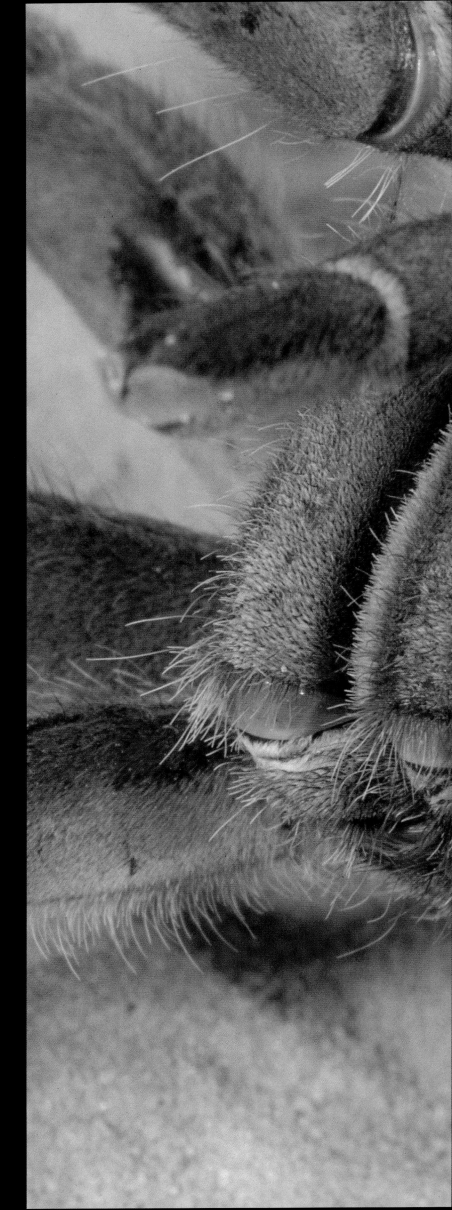

Early human cultures probably thought of venomous creatures as a kind of punishment meted out by evil gods. In the past, snakes and scorpions were undoubtedly feared more than spiders. But as snakes become less common, and scorpions are relatively localized in their distribution, only spiders can be described as the constant companions of humans. And the species of spider that adapt readily to man-made habitats tend to include the poisonous or, more correctly, venomous species [venomous means injecting venom; poisonous means toxic].

ALTHOUGH VIRTUALLY ALL SPIDERS POSSESS VENOM, ONLY A SMALL NUMBER OF SPECIES, PROBABLY FEWER THAN A HUNDRED IN THE WORLD, HAVE A SUFFICIENTLY POTENT AND EFFECTIVE BITE TO BE OF MEDICAL IMPORTANCE. SPIDERS USE VENOM TO QUICKLY IMMOBILIZE OR KILL THEIR PREY. IT IS ALSO USED IN DEFENCE AGAINST ANIMALS, INCLUDING MAN, BUT THIS IS ONLY A SECONDARY PURPOSE. THE VENOM IS DELIVERED VIA A BITE FROM THE CHELICERAE, OR JAWS.

Previous: King Baboon Tarantula with large fangs visible.

Below: Usambara Baboon Spider in threat position.

All spiders, with the exception of one family, the Uloboridae, possess venom glands. In most, the pair of glands lie in the head region, each one connected by a narrow duct to the fang, opening at a pore near the tip. Venom is ejected by contraction of the gland's musculature.

For humans, envenomation by a spider is usually less dangerous than that by bees, wasps or ants, which may attack in swarms and deliver many stings. A victim may feel the effects of a spider bite at its site (local), or throughout the body (systemic), or as a combination of both. The stings of bees, wasps and ants are usually only local in their effect unless the number of stings is very great.

Spiders' venoms are generally classed as either neurotoxic or cytotoxic. Neurotoxic venoms block the transmission of nerve impulses to the muscles, causing spasm and paralysis. When dealing with a big and dangerous insect, an injection which produces rapid paralysis, preventing escape or retaliation, is of great advantage to the spider. Cytotoxic venoms, on the other hand, work more slowly. They slow down the prey and at the same time begin the process of digesting it internally. In humans, such bites may cause tissue necrosis, leaving ulcers and scars which are slow to heal. Some cytotoxic venoms, for example those of the North American Recluse Spider (*Loxosceles reclusa*), may also be hemolytic, leading to kidney failure.

Spiders' venoms are complex mixtures of a variety of substances and each species has its own individual composition. For example, a particular compound in a neurotoxic venom may target and block specific channels in cell membranes, such as the calcium or sodium channels. Calcium channels have a role in cardiac and muscular function, whilst sodium channels are present in nerve and muscle cells – they are implicated in neurotransmission. Thus spider antivenin must be specific to a particular species. In recent years, scientists have been working to understand the composition and mode of action of some of these venoms. Their quest is to isolate pharmacologically-active compounds such as polypeptides, which may have a potential in therapeutic medicine, as in pain relief or the control of hypertension. Spider venoms have also been used in developing species-specific insecticides.

The black widow spider and its neurotoxic venom

The American black widow (*Latrodectus mactans*) is the best-known example of a spider with a neurotoxic venom. The effect of its highly toxic venom is to over-stimulate the neuro-muscular transmitters, acetylcholine and noradrenalin, causing paralysis of both the sympathetic and parasympathetic nervous systems. The combined effect is to impose a sudden and severe stress on the human body. The bite itself may not be noticed at first but serious pain is felt up to an hour later in the lymph nodes and spreads to the muscles, particularly around the abdomen. The face becomes covered in sweat and a painful grimace develops with swollen eyes (a condition known as *'facies latrodectismi'*). The brain is also affected and the victim may become delirious and obsessed with the thought that he will die. If the breathing muscles are affected by paralysis, it is possible that death will ensue. The best treatment is a combination of calcium gluconate, to quell the pain, together with an antivenin that neutralizes the toxin. Since the development in the 1930s of the black widow antivenin, the death rate has dropped below 1 per cent.

Black widow venom is claimed to be fifteen times more potent than that of a rattlesnake. But the quantity injected is minute and so mortality is considerably less than in the case of rattlesnakes. The principal toxic component of black widow venom is a protein called a-latrotoxin. Envenomation by a black widow usually affects a large part of the body and lasts for a number of days. Its effects depend on factors such as the quantity delivered, the rate of accumulation of toxin at the receptor site, and the process of de-activation via metabolism.

It is interesting to note that the effects of black widow venom differ across a range of animals. Tests have discovered that the cat is very susceptible to the venom, while dogs are quite resistant and high doses are necessary to cause death. Sheep and rabbits are almost entirely resistant. By contrast, some large animals are highly susceptible. Venom extracted

"IN THE MIDDLE AGES, AN ALLEGED EPIDEMIC OF SPIDER BITES GRIPPED THE TARANTO REGION OF ITALY IN A KIND OF MANIA FOR OVER 300 YEARS. THE PEOPLE CALLED THE LOCAL WOLF SPIDER THE TARANTULA AND THEY BELIEVED THE VENOM FROM ITS BITE COULD BE SWEATED OUT ONLY BY DOING THE TARANTELLA, A LIVELY DANCE. PROBABLY, IT WAS JUST AN EXCUSE FOR A PARTY! ANYWAY, THE SPIDER THEY BLAMED MAY NOT HAVE BEEN RESPONSIBLE. BITES WERE MORE LIKELY TO HAVE COME FROM THE LOCAL BLACK WIDOW ('MALMIGNATTE'). IN ITALY TODAY, THE DANCE IS STILL PERFORMED BUT THE NUMBER OF SPIDER BITES IS VERY FEW!"

from one black widow has caused the death of a horse, while the injection of a preparation from a macerated black widow has killed a large camel.

Species related to the American Black Widow include the Malmignatte of southern Europe (France), the Red-back of Australia, the Katipo of New Zealand, the Button Spider of South Africa, the Araña Brava of Chile, the Araña del Lino of Argentina, and the Araña Capulina of Mexico. They are among the thirty or so species that belong to the world-wide genus *Latrodectus*. Their geographic distribution tends to be in those parts of the world where grapes grow. None of them is larger than a thumbnail, and mostly their appearance (females) is globular, resembling a black grape, sometimes with red markings. The males are tiny and do not bite.

The recluse spider and its cytotoxic venom

The recluse spider (*Loxosceles reclusa*), common in North America, is much feared on account of its necrotic or cytotoxic venom. The necrosis it causes (loxoscelism) may heal slowly or not at all. Recluse spiders in general are delicate, small- to medium-sized, very ordinary-looking spiders, six-eyed, and brown in colour. Often they inhabit houses and, in their nocturnal movements, may crawl into clothes and bedding. Most victims are bitten while sleeping or dressing. The recluse spider in South America (*Loxosceles laeta*), which is highly venomous, is known as the Araña de los Rincones ('corner spider'), or the Araña de Detras de los Cuadros ('spider behind the pictures').

The first symptom following a bite, 20 per cent of which occur on the face, is usually a local swelling, accompanied by a burning-stinging sensation. The enzyme responsible for the necrosis is sphingomyelinase. Most patients also experience restlessness, vomiting and general malaise. There are two forms of loxoscelism: a cutaneous, non life-threatening form, and a dangerous, life-threatening form which affects the kidneys,

Above: *Brown recluse spider.*

Opposite: *Black widow spider (female). The red hourglass marking is an unmistakable feature for identification.*

Right: Six-eyed Sand Spider. Reputed to be one of the most venomous of all spiders but rarely encountered. Note sand grains adhering to the body.

causing the urine to turn black. In Peru, the mortality rate has reached as high as 10 per cent of all recluse spider bites. In the United States also, there are claims that deaths have occurred.

Worldwide, there are about 50 species of recluse spiders (Sicariidae) but the most important are: *Loxosceles laeta* (South and Central America), *Loxosceles gaucho* (Brazil), *Loxosceles reclusa* (southern USA), and *Loxosceles rufescens* (cosmopolitan, including USA, Japan, Australia and Europe). It is interesting to note that the category of venom they possess (cytotoxic) occurs also in some snakes (e.g. vipers).

Other spiders with cytotoxic venoms

The Six-eyed Sand Spider (*Sicarius hahnii*) is claimed to be among the most venomous of all spiders. It is a crab-like relative of the recluse spider. Laboratory assays of its venom have demonstrated that it is particu-larly potent. Recorded bites are lacking but the venom is likely to cause serious blood vessel leakage and tissue destruction, possibly leading to death. Fortunately, *Sicarius* is not commonly encountered as it lives in the more arid western half of southern Africa. It is also very shy and thus unlikely to bite humans. It buries itself in the sand and ambushes prey that wanders too close. Sand grains adhere to its abdomen, acting as a natural camouflage. If disturbed, it will run a short distance and bury itself again.

In South America, there are also other species with cytotoxic venom. In particular, the bite of some wolf spiders, for example *Lycosa raptoria* and *Lycosa erythrognatha*, relatives of the European *Lycosa tarantula*, are known to cause a deep, gangrenous necrosis. Three to four days following the bite, a swollen area, up to a foot across, usually develops into an ulcerous necrosis which later becomes a large permanent scar.

The bite of the slender sac spiders causes a characteristically yellow inflammation and sometimes a mild necrosis. These are quite small and pale, insignificant-looking spiders. Most regions of the world have at least one venomous example of the group: *Cheiracanthium japonicum* (Japan), *C. lawrencei* (South Africa), *C. diversum* (Australia), *C. punctorium* (Europe), *C. inclusum* (USA), and *C. mildei* (USA and Europe).

Two notorious species

While most spiders are not aggressive to humans, there are two notorious species, with highly toxic venoms, that stand their ground and will attack man if threatened. These are the funnel-web spiders of Australia (*Atrax* species) and the wandering spiders (*Phoneutria nigriventer* of southern South America and *Phoneutria fera* of northern South America). Both funnel-web spiders and wandering spiders may live in or around buildings, and both need to be treated with the greatest respect.

The Sydney Funnel-web Spider (*Atrax robustus*) is one of the world's most dangerous species. It is commonly found in the suburbs of Sydney and, quite unusually among spiders, it is the male that causes most of the bites. Drops of venom appear at the tips of the large fangs and the spider strikes repeatedly and furiously at anything that moves. The fangs can even penetrate the skull of a small animal or a human fingernail. If the fangs are embedded, it is extremely difficult to remove the spider, especially from a child's finger. Needless to say, the bite is immediately painful, partly because of the depth of penetration and partly because the venom is highly acidic.

Below Sydney Funnel-web Spider (Atrax robustus) threatening to bite.

The venom of *Atrax* is both neurotoxic and cytotoxic. The enzyme hyaluronidase breaks down tissue and assists penetration of the venom. Systemic effects begin within ten minutes: nausea, loss of sight, sweating, delirium, intense muscle twitching, abdominal cramps, an accumulation of fluid in the lungs, and paralysis of the breathing centre. The patient may sink into a coma with the accompanying risk of asphyxiation and cardiac arrest. In three cases of infant mortality, the victims died within 15 to 90 minutes of being bitten. Thus the prompt application of a tourniquet can be a life-saver. Adult victims, however, may take more than 30 hours to die. The venom has been studied in detail but the production of an antivenin has proved difficult. The noxious effects of the venom are seen only in man and monkey; other animals, such dogs, cats and rabbits, seem to be almost immune to it. The properties of the principal toxic component 'atraxotoxin' are somewhat puzzling but the poison is known to over-stimulate the sodium channels of the nerve and muscle membranes, causing eventual paralysis.

In parts of Brazil, 60 per cent of all spider bites are caused by the much feared Brazilian Wandering Spider (*Phoneutria nigriventer*). It is large, fast, agile and bites readily. When provoked it makes a threat display by raising the first two pairs of legs and exposing the red-coloured jaws. The bite of the spider causes immediate and intense pain which spreads throughout the whole body, causing quite a shock. The spider is aggressive to humans and, in contrast to other species, does not retreat when molested. It can leap onto, and climb rapidly up, the handle of a broom used to fend it off.

Wandering spiders have a legspan of about 12cm. They are common in urban areas of South America, where they find shelter and abundant food (cockroaches, other insects and small vertebrates). They live beneath fallen trees, in piles of wood, in the midst of building rubble, in banana bunches and bromeliads. They hunt actively

Below: Wandering spider. The jaws are reddish in colour and the spider is fast and agile.

at night, and rapidly overcome their prey with the strength of their venom; they do not construct webs. Often wandering spiders will enter houses, where they find hiding places during the day, such as in shoes and under furniture and doorknobs.

Other venomous spiders

Because of their appearance, tarantulas have always been imagined to be dangerous or even fatally venomous. But, despite their large size, they are much less dangerous than is commonly thought. Their poison glands are relatively small and lie entirely

within the chelicerae. It is likely the venom is intended only to assist in the digestion of prey. However, some genera, for example *Pamphobeteus, Acanthoscurria, Theraphosa and Phormictopus*, should be accorded respect because of the paralysis and necrosis they can cause in laboratory animals. In Africa, the large baboon spiders, *Stromatopelma, Heteroscodra, Pterinochilus, Hysterocrates* and *Phoneyusa*, are said to be capable of giving very painful bites. In India and Sri Lanka the most venomous genus is the beautifully-marked, arboreal *Poecilotheria*. In Australia, the most venomous

species is probably the Whistling or Barking Spider *Selenocosmia crassipes*.

Europe is one part of the world that enjoys relative freedom from venomous spiders. The widow spiders (*Latrodectus* species) are quite rare. In most of Europe, fatalities caused by spiders are virtually unknown. Nevertheless, the risk of meeting a venomous spider is increasing these days because of introductions from abroad. Significant bites are most likely to be caused by the Tube-web Spider *Segestria florentina* (now spreading from southern Europe), the tangle-web weaver *Steatoda nobilis* (now spreading from the Canary Islands), and the slender sac spiders of the genus *Cheiracanthium*. The cosmopolitan Wood-louse Spider (*Dysdera crocata*) is another European species that can bite. It is quite common in urban areas and easily recognized by its cream abdomen, red cephalothorax and legs. It has large jaws, which it uses to pierce the tough integument of its prey – woodlice. A bite on the leg from this spider may cause the limb to swell painfully for a period of two or three days, and dizziness can occur.

Even the smallest spiders can be troublesome. An improbable, but true, report of spider bites at a sewage farm in Birmingham, England, in 1974, involved thousands of tiny 'money spiders' (*Leptohoptrum robustum*). It had never been imagined that such spiders, little more than two millimetres in length, could bite through the skin and cause the slightest annoyance to man. But while maintaining the filter beds during the month of July, a number of workmen were bitten by swarms of these spiders, which dropped down their necks or crawled up their arms, causing inflammation and swelling. Subsequent investigation of the filter beds found incredibly high numbers of the tiny arachnids living among the clinker – about 10,000 per cubic metre – vastly more than their normal density in a natural habitat.

Above: *Powerful jaws of the Woodlouse Spider Dysdera crocata.*

Opposite: *African baboon spider.*

Venom – the conclusion

The venom of spiders is effective only if delivered into the bloodstream and not if taken into the digestive system. Most bites result from accidental contact, as spiders are usually not aggressive. For example, it must be emphasized that black widows are very timid; when disturbed, they often let themselves fall from their web and pretend to be dead. Usually they bite only when accidentally pressed against the body. Unfortunately, accidents can happen when gardening or dressing.

When it comes to finding and identifying the culprit, diagnosis is often virtually impossible. In many cases in which a spider has been blamed there is no specimen to confirm it. The question of identity is important, because different species of spiders have venoms with different chemical compositions causing, in turn, a range of clinical symptoms requiring different medical treatment. Furthermore, spiders tend to get the blame for assorted bites and stings which are actually caused by other creatures such as biting flies, bed bugs and fleas. However, the double puncture mark of a genuine spider bite should distinguish it from an insect's sting.

Many years ago, spider venoms were used by American Indians and the Bushmen of the Kalahari to paint their arrowheads. Today, researchers are working on spider venoms to isolate beneficial compounds for use in therapeutic medicine. For example, the partly digestive, necrotic venoms of the recluse spiders (*Loxosceles* species) are potentially useful in the dispersal of blood clots which may cause heart attacks. One remarkable discovery has shown that venom from a Caribbean tarantula (*Psalmopoeus cambridgei*) targets the same nerve cells (the 'capsaicin receptor') as the very hot and spicy chili peppers. Both plant and tarantula have thus evolved closely similar toxins to cause pain and deter enemies. The toxins are now being studied to gain an understanding of how the ion channels of the nervous system work. They give clues as to how blockers on these channels could be designed to treat persistent pain, for example from arthritis.

Some common myths

Myth no. 1:

The cosmopolitan Daddy-long-legs Spider, *Pholcus phalangioides* (sometimes erroneously called a 'harvestman'), has the most potent venom of all but is unable to inject it. *The venom may be potent but there are no records of bites by the species, so it cannot be established that the venom is the most potent (in human terms).*

Myth no. 2:

A colony of large, venomous spiders, of a species unknown to science, lives under Windsor Castle. *Following exaggerated claims in the media, sample specimens were identified by arachnologists as the European Cave spider (Meta menardi), a species not listed as venomous. Unfortunately, this update was not reported.*

Myth no. 3:

People have lost arms and legs because of spider bites. *In such cases, the limbs were lost by a gangrene-causing bacterium and not a spider venom.*

Myth no. 4:

Spiders can lay their eggs under human skin in wounds created by their bites. Other spiders come back to feed on the resulting wound and female spiders feed their babies from such wounds. *No truth whatsoever.*

Myth no. 5:

While sleeping, people swallow an average of four spiders a year. *A complete myth.*

Opposite: Daddy-long-legs Spider Pholcus phalangioides. *Its jaws (behind the palps) are very small.*

CHAPTER 8
SOCIAL SPIDERS

Most spiders are territorial loners who will eat
relatives and other spiders if they get the chance.
However, of the approximately 38,000 species of
spider that are known to science, a small number,
perhaps twenty species, are exceptions to the
norm. These are highly unusual spiders that are
able to live together harmoniously in social
groups. They can cooperate peacefully in web-
building, prey capture and nest maintenance.
They even care for their own young and, in
some cases, care for each other's.

WHEN CHARLES DARWIN VOYAGED TO SOUTH AMERICA, HE DISCOVERED A SPECIES (PROBABLY PARAWIXIA BISTRIATA) THAT ASSEMBLED IN HUGE COLONIES. IN THE BOOK OF THE VOYAGE, HE DESCRIBED THEM AS LARGE, BLACK SPIDERS, WITH RUBY-COLOURED MARKS ON THEIR BACKS. THEY BUILT VERTICAL WEBS SEPARATED FROM EACH OTHER BY A SPACE OF ABOUT TWO FEET BUT WERE CONNECTED BY COMMON LINES WHICH EXTENDED TO ALL PARTS OF THE COMMUNITY. DARWIN WAS OBVIOUSLY IMPRESSED BY THEIR LIFESTYLE, AND HE NOTED THAT, IN A KIND OF SPIDER THAT IS NORMALLY SO BLOODTHIRSTY AND SOLITARY THAT EVEN THE TWO SEXES ATTACK EACH OTHER, THE GREGARIOUS WAY OF LIFE IS VERY REMARKABLE.

Previous: Female social spiders with egg sacs.

In recent years, experts have found more examples of social spiders. The evolution of their social behaviour has become a particularly interesting subject. These spiders are interesting because of the comparison, not only with normal spiders, but also with social insects: ants, bees, wasps and termites. In contrast to the matriarchal societies of insects, social spiders have equality among individuals. They interact more like a pride of lions than like a hive of bees, where a queen rules hundreds of drones and thousands of workers. An individual spider could probably survive outside a colony whereas an isolated insect could not. In insect groups, workers are sterile and only the queen lays eggs,

whereas in a colony of spiders all individuals are able to reproduce. However, unlike insect societies, only one or two species of spider have reached the level of cooperative brood care.

The degree of gregariousness among spiders ranges all the way from loose aggregations, classed as parasocial, such as crowds of webs around a lamp, to the relatively few cases of genuine social life. The benefits of living together are thought to include safety in numbers, help in dealing with larger and stronger prey, and plenty of sexual partners, though fights among males often break out. Webs on the outside catch more insects, but those inside get protection from predators and receive 'early warning' vibrations. From the small number of known examples, it seems that social life among some spiders is remarkably successful. Individuals grow faster and they have a lower rate of mortality, so that fewer eggs are needed. Thus it is fascinating to see an organism one normally thinks of as antisocial, predatory and cannibalistic, forming large cooperative societies. Their communities can easily number many thousands of individuals.

Examples of social spiders

The tangle web-weaver, *Anelosimus eximius* (Theridiidae), is probably the most social of all the social spiders. It lives in colonies in the rainforests of South America where it builds a metre-long, hammock-shaped web suspended from the lush vegetation by long threads. The brown bodies of the spiders are about the size of a pea. They band together in colonies of hundreds to thousands, spinning their collective web above rivers

and roads, and where light filters through the tree canopy. Several generations of spiders live together in the community and, with constant repairs, the web can last for several years.

The adults of *Anelosimus* care for the young without making distinctions between their own progeny and those of others. This is extremely unusual among spiders. They guard the eggs against predators, move the egg cocoons to the parts of the web with the most comfortable temperatures, and feed the hatchlings. Working together, the spiders

Above: The web of tangle web-weaver Anelosimus eximius. *An old web that has collected many dead leaves.*

Opposite: .

An example of a web built by tangle web-weaver Anelosimus. *This metre-long web is populated by hundreds of individuals.*

can capture prey up to ten times their own size. They each throw sticky silk from their spinnerets to immobilize the prey and follow up with venom injected into the prey's leg joints. A spider on its own would be lucky to bag an insect twice as big as itself. When a colony grows too large, the city-like web starts to break up under its own weight. The spiders split into two or three groups, or the young females move away on bridges of silk to establish their own colonies.

The lace web weaver, *Mallos gregalis* (Dictynidae), of Central Mexico, is another of the most highly social spiders. Large numbers live together without territoriality, aggression or cannibalism, feeding simultaneously on prey caught in a large communal web. The web is a complex sheet of woolly silk extending over bushes and other vegetation. Within the web are sheltered chambers, which are probably designed to protect against the sun and the weather. Webs are constructed cooperatively and one spider may finish a task begun by another. When an insect becomes trapped in the web, several spiders approach, drag, pull and bite it, and then feed together. Many individuals, including adults of both sexes and immatures, may share the food. The people of Mexico call this spider *el mosquero* ('the fly trap') and they bring sections of its web into the home to protect against flies.

The African sheet web spider, *Agelena consociata* (Agelenidae), is also highly social. These spiders live in a huge communal web of up to 1,500 individuals, who catch prey and share it like lions at a kill. They never display aggression towards one another. The web consists of a network of horizontal sheets with silk lines above, which serve to knock prey onto the sheets. The webs often extend up into the forest canopy with the parts connected by vertical scaffolding. Each individual web encompasses some leaves and branches, held together by silk. Individuals can move freely between all parts. When prey falls onto a sheet, its movements attract the attention of all spiders in the vicinity. If it is large, several spiders attack it.

All individuals of *Agelena* share the reconstruction and prey-gathering tasks, although it is usually the smaller spiders that mend the web, while the larger spiders (generally females) deal with prey. The spiders are most actively engaged in web-building and maintenance activities at night, but they will capture and consume prey at any hour. When dealing with prey, the spiders communicate with each other via vibrations of the web fibres. Though cooperation occurs in the capture of larger prey items, fighting may occur between individuals who want to pull the prey in a particular direction. However, once captured, the prey is shared among the members of the colony.

The cribellate weaver, *Stegodyphus dumicola* (Eresidae), of Africa, is unusual among social spiders in that it lives in semi-arid regions, while most others inhabit jungle regions. It normally builds a web in a thorn tree, but will use a fence where trees are scarce. The web consists of numerous chambers and tunnels. The old remains of prey are built into the web to attract flies and other insects. When a prey item gets caught in the web, one or two spiders emerge from the nest within and a tug of war can ensue. If the prey item is too large for the spiders, they vibrate the web and reinforcements will arrive. Together, they can overpower the largest of praying mantids by their sheer numbers. Gabar Goshawks have been observed to take parts of these webs, complete with spiders, into their own nests. The spiders are said to catch the flies that irritate the young fledgelings.

Sub-social spiders

Some species of social spiders show a little more respect for each other's personal space. The colonial orb-weavers, *Metepeira incrassata* (Araneidae), which live in the mountains of Mexico, build interconnected

webs, but they do defend their own territory. Together, they lay long silk lines to which the individuals anchor their orbs. Colonies range in size from fewer than ten to several thousand individuals. Their webs span large spaces between trees on forest edges and in power lines along roadsides. Colonies have been known to stretch nearly two football fields in length and measure as much as four metres across and two metres high. Within the network, each spider defends itself from other spiders and predators, and deals with insects that land on its own orb. Insects that fly into the series of webs may bounce out of the first orb they hit, but are likely to bump into a neighbour's territory. On average, each spider captures about as much prey as its neighbours. *Metepeira* occurs year-round in this habitat, which is rich in prey, and reproduction is continuous, with overlapping generations. Although there is no parental care, the young usually stay within the colony.

Because their level of sociability is not high, colonial orb-weavers walk a fine line between the desire to share and the desire to be independent. When times are good, it is better to live in a large group, because there is plenty to eat. But when resources are scarce, each individual still captures about the *average* amount of prey, but that smaller amount may not be enough to sustain it. In such conditions, a spider's best chance would be to fend for itself. Theoretically, some would still starve, but others would survive.

The dome web spider, *Cyrtophora moluccensis* (Araneidae) in Southeast Asia and Australia, and *Cyrtophora citricola* in Africa and the Mediterranean, is not highly social in its behaviour but hundreds of individuals may build immense assemblages of webs. The masses can reach far up electricity pylons or almost completely cover trees. The webs are durable and each one occupies a space of up to a cubic metre, with a complex of vertical lines and a domed, trampoline-like sheet in the centre. The communities appear to be social but in fact individual spiders defend their webs and may attack intruding neighbours. However, death and injuries rarely occur. Young spiders are permitted to build their little webs within the framework of the adult's web.

The huntsman spider *Delena cancerides* (Sparass-idae) is widely distributed in Australia wherever there are suitable trees. It is a particularly unusual species and the only social huntsman known in the world. Its appearance is of a large, flattened, crab-like spider with a legspan that can reach as much as 14cm. Incredibly, these spiders are able to live together in colonies of up to 300 individuals, under the

"DARWIN WAS IMPRESSED WITH THE COLONIAL ORB-WEAVERS (PARAWIXIA BISTRIATA) THAT HE SAW IN BRAZIL. A TYPICAL COLONY NUMBERS A THOUSAND OR MORE WEBS INTERCONNECTED BY STRONG THREADS 10-15 M IN LENGTH. REBUILT AT SUNSET, THE INDIVIDUAL WEBS ARE USUALLY SMALL AND FINE-MESHED TO CATCH A STAPLE DIET OF LITTLE FLIES. RECENTLY, HOWEVER, IT HAS BEEN DISCOVERED THAT WHEN TERMITES ARE IN FLIGHT, AT WHATEVER TIME OF DAY, THE SPIDERS RESPOND BY SPINNING LARGER, WIDE-MESHED WEBS TO TARGET THESE BIGGER INSECTS MOST EFFICIENTLY."

Left: Young spiders in nursery-web – not to be confused with a web of social spiders.

peeling bark of a large tree. *Delena* is only one of two social spiders reported in the world that do not build a web.

The crab spider *Diaea ergandros* (Thomisidae), of Australia, is the other species in the world that builds no web. Again, it is a highly unusual species and the only social crab spider in the world. In Eucalyptus forests it constructs a nest out of long, flexible leaves, a wonder of arachnid architecture. Each year, the female *Diaea* builds the foundation for her nest with five or six leaves, then lays her egg cocoon in the inner chamber. She sits on the cocoon, like a mother hen incubating her eggs, and guards it against predators and parasites. When her 40 to 80 offspring grow strong enough, they tie more leaf layers around the nest. The spiders fold a leaf over and tie its ends together with silk. They then wrap another leaf around the first, and another, and another, so that the layered nest looks like a head of a cabbage. Many threads hold the communal leaf nest together. The nest serves as a labyrinth, in which the spiders can avoid invaders. It protects them from insects, birds and small mammals, as well as other spiders. Without this protective nest the crab spiders would probably be easy prey in the Australian forests. A family group grows up in this nest and works as a team to ambush bees, moths and butterflies that alight on or near the nest. However, unlike most social spiders, these fingernail-sized crab spiders live together for just one generation and then disperse to form the next season's nests.

Compared with most typical spiders, the mother *Diaea* invests heavily in each of her progeny. Normal, non-social crab spiders may lay ten egg cocoons per year, each containing as many as 1,000 eggs, and then leave them on their own. But *Diaea* lays just one cocoon, and she feeds the young throughout the year. When winter comes and food supplies grow scarce, she serves up her last meal to her young: herself. Scientists were surprised to find that the family groups will accept non-related but same-species crab spiders. The colony hunts with these outsiders and offers them asylum in the nest, which seems to be a violation of evolutionary rules. However, the apparent altruism does indeed serve a purpose. When food is scarce, the crab spiders eat these immigrants rather than their own siblings.

The origins of social behaviour

Social spiders have evolved separately in eight unrelated spider genera in Africa, the Middle East, the Americas and Australia. In fact, the repeated appearance of social behaviour has puzzled spider experts. Perhaps the answer is that communal living offers some spiders their only chance in a harsh world. For example, if leaving a nest is too dangerous, rebuilding a web each day is too demanding, and finding a mate is too difficult, then being sociable can offer a better chance of success than solitude. The early stages of social behaviour are seen in those species in which the mother cares for her young for a substantial period of time after hatching, such as the Maternal-social Spider (*Coelotes terrestris*). When they are sufficiently grown, the new generation of spiders disperse and make their own way in life. But, at some point in the evolution of the social species, it seems that the young just don't leave home.

Notwithstanding the huntsman and crab spider already mentioned, most social spiders are web-builders. The threads of their webs play a central role in communication between the individuals. All social species inhabit the tropics or subtropics. They tend to occur in wet regions, where large insects are abundant throughout the year. Because heavy tropical rains tend to break up the webs and nests, most of these species need to use relatively thick strands of silk; consequently, there is a large investment in silk for building and repairs. Cooperation is thus one way to make the tasks less onerous and expensive for a single spider, as well as allowing the community to catch bigger prey.

The benefits of social life are clear, but what about the costs? As the colony expands, it is likely to be found by parasites which will infest the egg cocoons. However, the most important consequence is that, over many generations, it inevitably leads to inbreeding. Within a single colony, the individuals may be genetically identical. If life is relatively easy, it can be acceptable to be a clone but, without genetic variation, a population is at risk of being wiped out by an epidemic of disease. For example, a mysterious disease of spiders swept through Panama in 1983, killing entire colonies of social spiders.

Since mating with close relatives is likely to reduce fitness, most animal species disperse to avoid inbreeding. But social spiders

do not disperse and yet somehow they manage to tolerate repeated inbreeding. Undoubtedly their strategy is risky but social spiders enjoy some considerable advantages. They avoid the need for dispersal outside the community and this means less exposure to predators. Also they do not risk failure in trying to locate a mate.

The evolution of social spiders

The original theory of evolution considered natural selection to operate on an individual level. However, the idea of selection on a group level may provide insight into how social spiders have evolved. According to this concept, certain behaviours benefit entire communities, or species, of animals rather than individuals. Male deer, for example,

compete with each other through non-lethal displays. This type of behaviour may have evolved because it led to fewer deaths for the species as a whole, rather than to breeding advantages for the individual.

Today, biologists consider evolution at the level of whatever carries a gene. In social spiders, an entire inbred group may be the vehicle that carries a gene. The sex ratio among social spiders supports this theory. Often, 90 per cent of a population is female, a sex ratio that appears to benefit colonies in their competition with other colonies. The more fertile females there are in a nest, the faster the colony grows to a safe, productive size, and the more new colonies the group can establish. Males are relatively less important.

Above: Group of social spiders (Stegodyphus) attack grasshopper prey as a cheeky jumping spider comes close (left).

CHAPTER 9
SPIDERS AND HUMANS

Depending on your point of view, spiders can be a source of fear or fascination. Perhaps there is no other animal that evokes such a mix of emotions at the same time; we are fascinated by spiders because they seem both beautiful and dangerous. The more you learn about them, the more you will appreciate spiders, rather than fear them. In classical times, and probably long before, it was known that spiders' webs were excellent for healing wounds. This ability of silk appears to come from its remarkable combination of softness and strength, its immunity from microbial attack and, reputedly, its power to coagulate the blood. Today, the challenge is to use the industriousness of spiders to create materials stronger than those produced so far by human beings. By emulating some of the spider's web-spinning abilities, engineers can improve the production of materials to enhance the performance of many products, from tennis rackets to bulletproof vests.

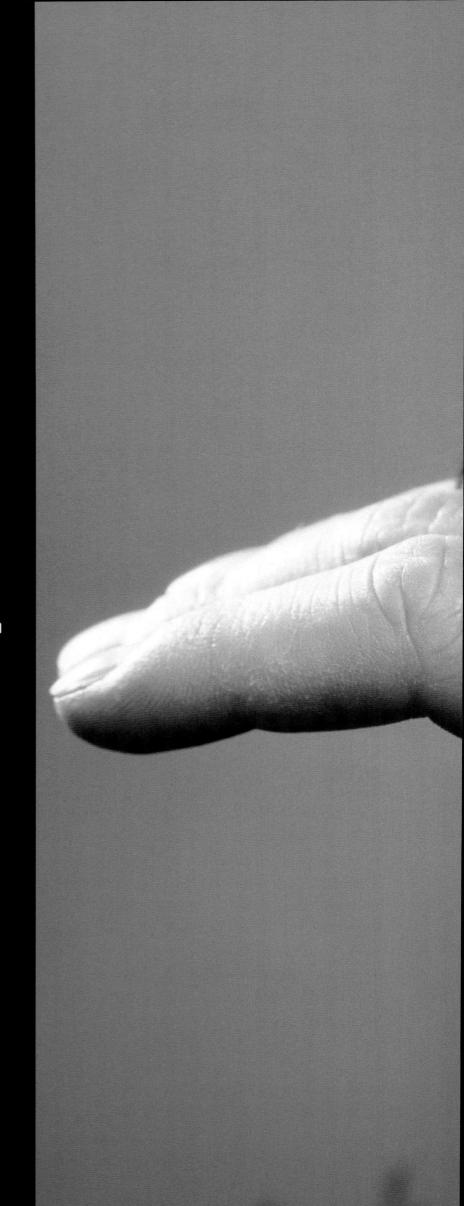

HISTORY TELLS THAT, AFTER A DEFEAT BY KING EDWARD OF ENGLAND IN 1306, KING ROBERT THE BRUCE OF SCOTLAND WAS GREATLY INSPIRED BY A SPIDER WHICH TRIED REPEATEDLY TO ANCHOR THE FIRST LINE OF A WEB. ITS PERSEVERANCE SHOWED BRUCE THAT "IF AT FIRST YOU DON'T SUCCEED, TRY, TRY AGAIN". HE EMERGED FROM HIDING, RALLIED HIS MEN, AND IN ONE LAST ATTEMPT DEFEATED EDWARD'S ARMY IN 1314. SPIDERS CAPTURE OUR IMAGINATION BECAUSE THEY ARE SMALL AND YET SO SKILFUL. FEAR MAY EXIST BUT IT IS SURELY OUTWEIGHED BY OUR ADMIRATION.

Previous: *Tarantula spider on hand.*

Below: *Haitian Brown Tarantula (*Phormictopus cancerides) *from the islands of Hispaniola and Puerto Rico.*

ARACHNOPHOBIA

Arachnophobia, or the fear of spiders, is probably more common than the fear of dentists and second only to the fear of heights. It is the most prevalent of all animal phobias and can afflict almost anybody. One sufferer confessed: "I couldn't even write the word spider. I daren't put my bag on the floor in case a spider crawled into it and I couldn't go into a room until someone else had checked that there were no spiders inside".

A phobia is an anxiety or fear that is out of proportion to the danger of the situation. Its symptoms are similar to those of a genuine fear response: sudden apprehension, loss of control, shortness of breath, increased heart rate, faintness, sweaty palms and trembling. In extreme cases the individual may become

paralysed. But many arachnophobic people do in fact recognize that there is no need to fear spiders. Often their phobia is not a fear of what a spider may do, but simply a loathing of the sight of it. Such people would readily admit that the possession of eight long legs, a hairy body and scuttling movements, particularly in darkness or inaccessible corners, awakes in them some primitive, probably innate, horror and repulsion that no argument can overcome.

Can arachnophobia be explained?

How can we explain the occurrence of arachnophobia? Is the fear learned or acquired, or is it an innate instinct, genetically determined? Is there an inherited tendency within a family to be anxious or nervous, and to have such fears? Over the years there have been many theories, but they remain unproven because there are such variations in individual experiences. Some assert that arachnophobia is a by-product of modern, urban life, with its separation from nature and natural hazards. But the reality is that the fear of spiders does have a long history. Two thousand years ago 'a dreadful plague of spiders' caused the depopulation of lands in Abyssinia (Ethiopia) and, in southern Europe during the Middle Ages, spiders were blamed for long episodes of mass hysteria known as Tarantism.

It is probably true that arachnophobia often starts in childhood. The very youngest children are usually not afraid of spiders but, as they get older, the number of children with such fears tends to increase. When arachnophobia persists into adulthood it tends to worsen and rarely disappears without treatment. Anxiety is certainly part of the problem, but which ingredient comes first? Is it the tendency towards anxiety? Or is it that spiders actually make a person anxious? There is a theory that arachnophobes are more than usually repelled by dirt and creatures such as slugs, cockroaches and maggots. It is probably true that a strong dislike of insects, or entomophobia, is often related to an obsessional desire for cleanliness. According to this 'unclean' theory, arachnophobia may owe its origin to medieval Europe where the presence of spiders was associated with households infected by bubonic plague. The homes of the dead and dying fell into neglect and spiders moved in, acting as tell-tale signs of infection. Furthermore, the memory of this is supposed to have passed on from gener-ation to gene-ration. As a result, sufferers are likely to have a phobic relative. Indeed, some experts do believe in a tendency to inherit the fear.

Overcoming arachnophobia

Cognitive-behavioural therapy (CBT) is the standard clinical treatment for arachnophobia. It attempts to overturn a way of thinking that has become stuck in a rut. Usually it involves some sort of exposure to spiders combined with ample moral support. The exposure is progressive and works towards a gradual desensitization using a series of spiders which are increasingly difficult for the patient to deal with.

The best form of self-help is to learn about spiders. Arachnophobia is largely a fear of the unknown. If one talks to an arachnophobic person, it soon becomes clear that their many misapprehensions have caused them to suffer for years. Exaggerated beliefs about spiders are often dearly held but mistaken. Thus if people can learn about spiders, based on a true understanding of their behaviour and capabilities, the problem of arachnophobia will begin to disappear.

Arachnophobia – the conclusion

Clearly, a degree of fear is normal. Most people are afraid to touch spiders or allow them in their homes. Also it is probable that phobias of spiders (and snakes) tend to manifest more frequently than other phobias. But many types of spider, for example jumping spiders and money spiders, usually cause no fear to anybody. Probably, few people realise that the incidence of arachnophobia in a particular country is very much influenced by the presence or absence of the types of spider that are perceived as 'creepy'. For example, in England, where the abundant, long-legged house spiders (*Tegenaria* species) are public enemy number one, arachnophobia is very common. But in New England (USA), spiders have a much lower profile and are rated well below cockroaches in terms of public distaste. It all depends on your local spiders.

Traditionally, the psychoanalytical profession has regarded phobias, such as those of spiders, as acquired. Experts have viewed patients as having learned to be afraid of spiders, and treatment has been based on the principle that what is learned can be unlearned. But personal accounts from sufferers suggest that natural instinct plays an essential role, and that phobias are simply exaggerated extensions of these innate, instinctive fears. It may seem counter-intuitive, but the evidence suggests that the fear of spiders is not learned; it is innate. Indeed, it is the state of *not being afraid of spiders* that is learned. An arachnophobic person is someone who has not yet learned to suppress, or overcome, the fear.

BUILDING SPIDERS' WEBS IN OUTER SPACE

In 1973, a space station called Skylab 3 was launched by NASA to conduct experiments in zero gravity, outside the Earth's atmosphere. Two female Garden Spiders (*Araneus diadematus*) were included on board to see how they performed in conditions of near-weightlessness.

A week after lift-off, spider number one was released by an astronaut into an experiment cage where it could build a web. Spider number two was kept in a small tube until

Below: *House spider* (Tegenaria) *in the funnel retreat of a typical, dusty web.*

released into the cage four weeks after launch. A camera was positioned to take pictures of the activities of the spiders and their webs. On transfer to the cage, both spiders made what have been described as 'erratic swimming motions'. They needed time to adapt to near-weightlessness, but after one day in the cage, spider number one produced her first rudimentary web in a corner. And the next day, she built a complete web. It was a great success in demonstrating the remarkable adaptability of the spider.

The experiment succeeded in showing that spiders can build orb webs in zero gravity. While the two spiders had considerable difficulty in adjusting to the strange conditions, it was interesting that their work differed only in minor details from Earth webs. The silk spun in space varied in thickness, whereas on Earth it has a uniform width. And curiously, the webs in space were perfectly symmetric, while on Earth they are usually asymmetric, with a larger lower half containing more radii than the upper. Also, in space the spiral thread went round without any interruption – there were few turning points. These differences in construction were attributed to the absence of gravity as a cue. On Earth, spiders are able to use gravity (via their body weight) to adjust the strength of the silk. Silk strength is highest when it is pulled rapidly from the spinnerets, as when the spider drops down, for example. Understanding the way a spider makes up for the lack of gravity when building its web in space may therefore assist engineers to create stronger construction materials.

SPIDERS AS AN INSPIRATION TO BUILDERS AND ARCHITECTS

Supporting structures in buildings experience either a pull (tension) or a push

Above: *Garden spider (Araneus diadematus) sitting at the hub of it's web.*

Right: The Millennium Dome, London.

(compression). Supports under tension, such as cables, are made from material that can stretch to some degree. These supports are usually thin and rope-like. On the other hand, structures under compression, such as columns, are made with solid materials and tend to be thick. Loads, such as a person walking across a bridge, or an insect hitting a spider's web, subject the structure to forces.

Structures such as bridges and spider webs are designed to distribute stresses efficiently across their length and width. As builders, spiders have attracted the attention of architects and engineers. Their webs are dynamic, built for movement and designed to absorb energy. Man's structures by contrast are usually fixed. Thus in order to protect against earthquakes, for example, we need to study the building techniques of spiders. According to the arachnologist, Fritz Vollrath: "Spiders are the ultimate architects of lightweight structures, as well as the manufacturers of superfibres. After millions of years of evolution, they are masters of cost-sensitive engineering and wizards of polymer science. The spider's tricks are always elegant and functional because nature rewards only success".

Because of its sheer scale, the London Millennium Dome was hailed as the engineering breakthrough of its decade. But a tent-weaving spider (*Oecobius* species) was already building dome webs on the same principle, i.e. a tent under tension, in miniature. The web is built upside down from the underside of a rock. It uses the weight of a number of small stones on silk lines to supply the stays that, in the Millennium Dome, are provided by complex steel poles. The web is not just a shelter but also a trap, because insects entering are caught in the woolly silk.

Synthesizing spider silk

Scientists have long envied the strength, lightness and elasticity of spiders' silk, but have been unable to synthesize it. The silk

Right: *Inside the Millennium Dome, London.*

possesses a rare combination of strength and toughness, meaning that not only can it hold relatively heavy objects, it can also stretch to great lengths without snapping. Normally we can make material very strong but at the expense of toughness. And we can make things very tough but at the expense of strength. Combining the two characteristics, as the spider does, is the challenge.

Historically, there have been many attempts to milk spiders for their silk but employing them in any numbers is not practical, because they are cannibals. Now, modern methods of biotechnology have enabled us to replicate spider silk in the laboratory using genetic engineering techniques. The science of bioengineered materials is a growing area of research involving the creation of synthetic proteins based on those in nature, to make materials with advanced properties, such as lightweight bullet-proof vests, medical sutures and artificial ligaments.

Of all the different types of silk, the dragline of the Golden Orb Spider is the very strongest fibre. It is actually stronger than the strongest synthetic fibre – the bullet-proof Kevlar. Whereas Kevlar can stretch up to 4 per cent before breaking, spider silk will stretch as much as 15 per cent before breaking. In nature, silk strands elongate in conditions of a sudden load, for example when an insect is caught, and in effect turn the prey's momentum into heat. The strands then rebound gently so as not to catapult the insect back out. This ability to dissipate energy would make silk ideal for bullet-proof vests.

The objective is to produce the proteins of spider silk in large quantities. Probably this will be done in an industrial process, by transferring the specific, silk-making genes of spiders to bacteria, such as *Escherichia coli*, which can then be reproduced as a cloned colony in large vats, to mass-produce the desired protein. A possible alternative is to replace the genes of silkworm caterpillars with those of spiders.

Another possibility is that animals could mass-produce the material more efficiently than vats and machines. In the USA, genes from an orb-weaver have been inserted into the genome of a Nigerian goat (called Charlotte's Goat). The goat's mammary glands are then able to reproduce the complex proteins that make up spider's silk. The goat's milk looks no different, but when the proteins are filtered and purified into a fine white powder, they can be spun into a tough thread similar to that of spiders.

Compared with human technology in creating fibre materials, spider silk is a much greater success. It is made at ambient temperatures, at near ambient pressures, and uses only water as a solvent. Furthermore, it seems to cost the spider so little energy that its manufacturing expenditures hardly show up in terms of its daily metabolic expenses. Thus spider silk, which is similar in molecular structure to that of the silkworm, but much stronger, is a material that could not be friendlier to our environment.

SPIDER CONSERVATION

Why do we need to conserve spiders? Primarily, because they are as important in the balance of nature as any other group of animals. They provide food for birds and many other types of wildlife, and also serve a very useful function as far as humans are concerned through their consumption of vast numbers of pest insects. Pesticides have reduced spider numbers and yet they are among the most useful allies of farmers and growers. Yet even if they were not useful, spiders would still deserve our protection as one of the most diverse and fascinating groups in the natural world.

The Chinese may have been the first to conserve spiders. Next to the country's paddy-fields, little straw huts were traditionally built by farmers to house spiders during the winter. By being protected in this way, the spiders survived the cold and were ready in the spring to attack the many crop-damaging insects. More recently, spiders have also been encouraged in a traditional Belgian beer brewery to play an environmentally friendly role in the control of fruit flies (*Drosophila* species). Around the barrels in the brewery, large numbers of spiders' webs keep the flies away so that the unwanted microorganisms that the latter carry cannot affect the fermentation process and thereby spoil the beer.

While there may be some exceptional examples of human appreciation of spiders, such as cultivating them in breweries, in general they have received very little public support. They have also been largely forgotten by the conservation community.

Nevertheless, by stressing the value, beauty and interesting behaviours of spiders, as this book has tried to do, it is hoped that attitudes towards these invertebrates may change. As more information becomes available, we are gaining a better understanding of the role of spiders in ecology and of the need to improve their protection. At this point, however, it must be admitted that we know so little about the numbers and distribution of species in the world, that it is difficult to know which are threatened and which may be already extinct. Currently, the 2006 Red List of the International Union for the Conservation of Nature (IUCN) classifies 13 species of spiders as *Threatened*. In reality, the true number is doubtless much greater.

Below: *Mexican Red-kneed Tarantula.*

The threats to the survival of spiders are many. Numerous species are restricted to particular habitats, such as islands, caves, forests, heaths and wetlands. Such habitats can easily be lost through urbanization, mining, drainage, and agriculture. Spiders are also threatened by pollution, acid rain, climate change and the introduction of alien species such as fire ants (*Solenopsis* species). Vandalism is another danger; at the Tooth Cave in Texas, where the tiny (1.6mm), blind spider *Neoleptoneta myopica* is found, a locked gate has been installed. Furthermore, over-collecting for the pet trade is threatening some of the large tarantulas, such as the beautiful Mexican Red-knee Tarantula (*Brachypelma smithi*). This species is currently on Appendix II of CITES (the Convention on International Trade in Endangered Species), which requires a permit to be issued by Mexico for every spider exported to a CITES-observing country. This species is also the subject of a captive-breeding programme pioneered by London Zoo (Invertebrate Conservation Centre).

As with most conservation, spider conservation is essentially about the protection and maintenance of habitat. This can be as simple as leaving parts of the lawn unmowed, or setting aside strips of weedy cover around fields and plantations from which spiders can move out into the crops. For example, a density of 20 wolf spiders (*Pardosa ramulosa*) per square metre can reduce leafhopper insects in American rice fields by up to 90 per cent, as each spider consumes about 5–15 insects per year. But spiders are not the greatest of biological control agents because they kill beneficial as well as pest insects. However, their ability to exist in large numbers, on relatively little food, throughout the year, gives them a useful role in regulating the fluctuations of insect numbers. Thus, rather than controlling specific insects, spiders have a general buffer effect, keeping most insects under

Above: *Great Raft Spider (Dolomedes plantarius). The pride of two sites in the U.K.*

control while supplying food for other animals and thus maintaining the flow of life.

Worldwide, there are only a handful of examples of habitats which have been protected to preserve a rare and particular species of spider. In the United States, the prime examples are the Hawaiian lava-tube habitats, home to the No-eyed Big-eyed Wolf Spider (*Adelocosa anops*), and the Tooth Cave in Texas where *Neoleptoneta myopica* occurs. Both are protected by the U.S. Endangered Species Act. In the United Kingdom, the Great Raft Spider (*Dolomedes plantarius*), discovered in 1956 at Redgrave and Lopham Fen in East Anglia, has been given full protection under the Wildlife and Countryside Act, 1981. It occurs widely in the Palaearctic region, but is restricted to just two sites in the UK, where it is classified as *Endangered*. It is a relatively large (up to 23mm body length), semi-aquatic spider which lives in wetlands. However, in spite of conservation measures, falling water levels on the fen have caused the population to diminish in recent decades and in most years it probably numbers little more than 100 adult females; it is therefore highly vulnerable to extinction. In 1988, the species was discovered at a second site, the Pevensey Levels in southeast England, where a bigger population has been estimated at 3,000 adult females. Probably it had passed unnoticed there for so long because the species resembles the relatively common *Dolomedes fimbriatus*.

A major challenge in the conservation of spiders is the limited and highly dispersed nature of the available data. Information on species, habitats and countries is buried in hundreds of technical taxonomic papers, much of which is unobtainable or rare and now outdated or unreliable. In most countries, there is a serious need for the creation of identification keys which would allow non-specialists to conduct spider surveys. Unfortunately, little funding is currently available to do this kind of work and yet without such information species in need of conservation risk being overlooked. In the future, it is hoped that changing attitudes and a greater interest in spiders will bring more effective results. Improvements in conservation will also require more cooperation and communication between spider enthusiasts and the conservation community. In all countries, if not already created, it will be worth setting up a spider database, to include all known information on local species. This would serve as an essential tool for the conservation management of spiders and go a considerable way to helping ensure greater understanding of their fascinating behaviour and contribution to the wider ecosystem.

"WE SHOULD SUPPORT OUR LOCAL SPIDERS. SPIDERS FACE MANY DIFFICULTIES. THEIR BODIES ARE NOT PERFECT. THEIR CIRCULATION DOES NOT DELIVER OXYGEN EFFICIENTLY AND YET BLOOD PRESSURE MUST BE MAINTAINED TO KEEP THE LEGS WORKING. EACH TIME A NEW EXOSKELETON HARDENS, THE SPIDER CAN'T EVEN STAND, LET ALONE DEFEND ITSELF. FOOD MUST BE PRE-DIGESTED AND SUCKED IN AS A SOUP. MANY HAVE FEEBLE EYES. AND WHEN IT COMES TO REPRODUCTION, A SPIDER MAY LAY HUNDREDS OF EGGS, BUT FEW WILL LIVE TO ADULTHOOD."

GLOSSARY

Abdomen The rear of the two main parts of a spider's body.

Accessory sex organs The palps of male spiders, which function as intermediate sex organs to transfer sperm from the male's genital opening to that of the female.

A-latrotoxin The principal toxic component of Black widow spider venom.

Antennae A pair of thread-like 'feelers' on the head of many arthropods but not spiders.

Antivenin A specific antidote to a venom delivered by a venomous animal.

Arachnida A class of arthropods characterised by the possession of eight legs. Includes spiders, scorpions, mites, ticks, pseudoscorpions and harvestmen.

Araneomorphae The largest and most advanced of the three divisions (suborders) of spiders. See also Mygalomorphae and Liphistiomorphae.

Arthropoda The largest group (phylum) in the animal kingdom, defined by jointed legs and tough exoskeleton. Arthropods include insects, arachnids and crustaceans.

Atraxotoxin The principal toxic component of Sydney Funnel-web Spider venom.

Attack behaviour The final approach by a spider towards it's prey.

Ballooning Dispersal of small spiders by means of airborne lines of silk.

Biotechnology The science of reproducing materials from nature by basing them on biological systems.

Book lung(s) Respiratory organ consisting of a cavity in which air is brought into contact with blood-filled layers. Book lungs appear as pale squares on the underside of the abdomen. Mygalomorphs have two pairs while araneomorphs usually have only one pair (plus a system of tracheae).

Calamistrum A comb-like series of bristles on the last but one segment of the fourth leg in cribellate spiders.

Capture spiral See Sticky spiral

Carapace The shell-like covering of the first part of a spider's body (the cephalothorax).

Cephalothorax The first of the two main parts of the body of a spider; a single structure formed by the fusion of the head and thorax.

Chelicerae The technical term for the spiders jaws. Each chelicera comprises a basal part and an articulating fang.

Class A unit of classification. A class (e.g. Arachnida) consists of a number of related orders. A number of classes are grouped into a phylum (e.g. Arthropoda).

Cocoon A silken brood chamber made to protect the eggs of a spider.

Cohabitation Spiders are normally solitary, but occasionally the male and female may be found in a single cell, e.g. in sac spiders prior to the female becoming adult.

Courtship ritual The specific behaviour which occurs in the build up to mating.

Coxa The first segment of a spider's leg (nearest the body).

Cribellate spider See Cribellum and Calamistrum

Cribellum An organ which spins silk of an extremely fine and woolly quality. The cribellum appears as a sieve-like plate and is present only in 'cribellate' spiders.

Dragline A line of silk attached for safety to the substrate. A spider may quickly descend on a dragline and climb up again later.

Electrostatic force The force of attraction produced by a stationary electric charge.

Epigyne The external structure associated with the reproductive openings of adult females of most spider species.

Exoskeleton The tough, external covering of the body of all arthropods, including spiders.

Eyes
(compound) Compound eyes with many facets occur in insects and crustaceans but not in spiders.
(posterior median) The two central eyes in the second row.
(principal) The two central eyes in the first row, usually forward-facing.
(simple) The eyes are classed as 'simple' but some spders have lenses capable of forming clear images.

Family A unit of classification. A family (e.g. Salticidae) consists of a number of related genera. Approximately 110 families make up the order Araneae (spiders).

Fang The needle sharp, claw-like part of each chelicera, through which the venom is injected into the prey.

Femur The third and often largest segment of a spider's leg

Fossil (living) An extant species considered to be exceptionally ancient and primitive.

Frequency range A range of sound waves measured in hertz, e.g. 20–20,000 Hz.

Genus A unit of classification. A genus (e.g. Salticus) consists of a number of related species, all sharing the same first name. Related genera are grouped into families.

Gossamer Long lines of spider silk. Where abundant, the lines may mass together to form a light film or sheet.

Harvestman A member of the Arachnida, related to spiders but lacking venom glands and possessing a one-part body, not two.

Hub The central patch of an orb web, like a bull's eye, where the spider often sits.

Hydraulic extension Straightening of the legs due to the internal pressure of body fluid.

Iridescence Metallic colours that alter when the angle of view is changed.

Kleptoparasite An animal that steals the food of another, e.g. a spider that feeds in the web of another species.

Liphistiomorphae The most primitive of the three divisions (suborders) of spiders. See also Araneomorphae and Mygalomorphae.

Mating plug A physical barrier in a female's genital tract which prevents insemination.

Matriarchal society A hierarchical society headed by a female, e.g. a colony of bees.

Metabolic rate The rate of metabolism (burning chemical energy from food) measured by an animal's consumption of oxygen.

Metatarsus The sixth segment of a spider's leg.

Molecular weight The mass per molecule (total atomic weight) of a compound.

Moulting The skin-changing process which occurs 5–10 times in the life of a spider.

Mygalomorphae One of the 3 divisions (suborders) of spiders; considered to be relatively primitive compared with Araneomorphae. See also Liphistiomorphae.

Neurotransmitter A locally-acting chemical compound, e.g. acetylcholine, that transmits an impulse across a synapse, i.e. nerve to nerve or nerve to muscle.

Orb web A two-dimensional web, roughly circular in design. Radial threads, like the spokes of a wheel, are crossed by a spiral thread which circles round and round.

Order A unit of classification. An order (e.g. Araneae) consists of a number of related families. Orders are grouped into classes (e.g. Arachnida).

Palps (or pedipalps) The two leg-like feelers arising just in front of the legs. In adult males the palps are modified for the transfer of sperm.

Palpal organ The more or less complex structure on the end of the adult male palp. See also Accessory sex organs.

Parasite An organism which spends part or all of its life on or in another species (the host), taking food from it but giving nothing in return.

Parasocial A community of individuals that live together but do not share food.

Patella The fourth, knee-like segment of a spider's leg.

Pedicel The narrow stalk connecting the cephalothorax to the abdomen.

Percussive sounds Noises produced by striking two objects together.

Pheromone A chemical secreted in minute quantities which brings about a response in the opposite sex.

Polymer(s) Materials (natural and synthetic) built of repeated molecular units, e.g. rubber, proteins, starch, cellulose, nylon and polythene.

Radius One of a number of silk lines radiating out from the middle of an orb-web, like the spokes of a wheel.

Retreat A spider's shelter, usually made of silk.

Sexual dimorphism The marked difference in appearance between males and females of the same species.

Silk A natural fibre, indispensable to spiders.
  (cribellate) Extremely fine, woolly silk produced by cribellate spiders. Also known as hackled band silk.
  (filaments) Fine threads of silk that emerge from the spinnerets.

Silk glands Abdominal glands (up to 6 types) producing silk in liquid form.
  (aciniform) For the swathing band and sperm web.
  (aggregate) For the spiral thread (in orb-weavers).
  (ampullate) For draglines and frame threads.
  (flagelliform) For sticky silk (in orb-weavers).
  (piriform) For the attachment disc of draglines.
  (tubuliform) To construct cocoons.

Social behaviour Cooperative interactions between the members of a society.

Species The basic unit of biological classification. Ideally, a species is defined as a group of organisms that interbreed to produce fertile offspring. A number of closely related species may be grouped together in a genus.

Spermathecae Paired sacs or cavities that receive and store sperm in adult female spiders.

Spiderling A newly hatched spider.

Spigots Microscopic nozzles, at the apex of a spinneret, through which emerge fine filaments of silk.

Spinneret(s) Appendages at the end of the abdomen,

numbering up to 3 pairs and connected by ducts to the silk glands. See also Spigots.

Spines A thick, stiff hair or bristle.

Spiracle An opening to the tracheae, on the underside of the abdomen.

Spurs Thorn-like projections on the legs and palps.

Stabilimentum A band of white silk usually in the form of a zigzag near the hub of an orb web. Stabilimenta are characteristic of spiders belonging to the genus Argiope.

Sternum The oval plate which occupies the space between the legs on the underside of a spider's body.

Sticky beads/droplets/globules Beads of sticky silk on capture threads for increased efficiency in catching prey, particularly in the families Araneidae and Theridiidae.

Sticky spiral The thread that circles across the radii of an orb web and usually beaded with droplets of sticky silk. Also known as the capture spiral.

Stridulating organ A file and scraper system to make sound; variously located on chelicerae, palps, legs, abdomen and/or carapace.

Substrate (substratum) The surface or medium to which an organism is attached.

Surface tension The elastic, skin-like property on the surface of a liquid.

Swathing band A thick band of silk used to wrap dead or living prey.

Symmetric (web) A symmetric orb web is roughly similar above and below the hub.

Tarantula Originally the name applied to the European wolf spider *Lycosa tarantula*, but now used for the large myga-lomorph spiders (family Theraphosidae) of tropical regions.

Tarsus The seventh segment of a spider's leg (farthest from the body).

Thread(s)
  (anchor) Guy lines that hold the frame threads of an orb web in place.
  (bridge) An important thread which crosses a gap and supports an orb web.
  (capture) The parts of a web that are important in catching prey.
  (frame) Semi-permanent lines surrounding an orb.
  (mating) A line attached by a male to a female's web during courtship.
  (permanent spiral) The spiral laid on return to the hub replacing the temporary one.
  (signal) A line of communication between the web's hub and the spider.

(support) Lines that support a web from above.
(temporary spiral) The initial spiral laid from hub to periphery of an orb web.

Tibia The fifth segment of a spider's leg.

Tracheae Fine tubes allowing air from the respiratory openings to reach the body tissues.

Trip-line(s) or trip-thread(s) Silk lines extending from the opening of a spider's burrow to transmit vibrations from passing insects.

Trochanter The short, second segment of a spider's leg.

Tubercle(s) Thorn- or wart-like irregularities on the surface of the body.

Urticaria In the case of spiders (New World tarantulas), urticaria (pain and inflammation) may be caused by contact with their irritant hairs.

Venom Poisonous fluid of animal origin containing active compounds (toxins).
  (cytotoxic) Causing damage to the body tissues.
  (hemolytic/haemolytic) Causing kidney failure.
  (necrotic) Causing the death of tissue.
  (neurotoxic) Affecting the nervous system.

Venomous Capable of injecting a venom by means of a bite or sting.

Web
  (barrier) A tangle of threads sheltering a spider at the hub of an orb web.
  (decorated) An orb web with stabilimenta (usually zigzags).
  (dome) A complex structure with central dome.
  (funnel) A web which funnels into a retreat at the rear.
  (lace) A web of woolly or frilly appearance made by a cribellate spider.
  (ladder) An asymmetric orb web with a large extension above or below the hub.
  (nursery) A web built for protection of the young.
  (orb) See Orb web.
  (reduced) A web made of a few threads or even a single thread.
  (sheet) A horizontal web with a more or less sheet-like surface.
  (sperm) A tiny platform onto which a few drops of sperm are extruded.
  (tangle) A web with taught capture threads above and below a central layer.
  (triangle) A web shaped as a triangle, with an anchor thread held by the spider.
  (tube) A web built into a crevice with trip threads radiating out.

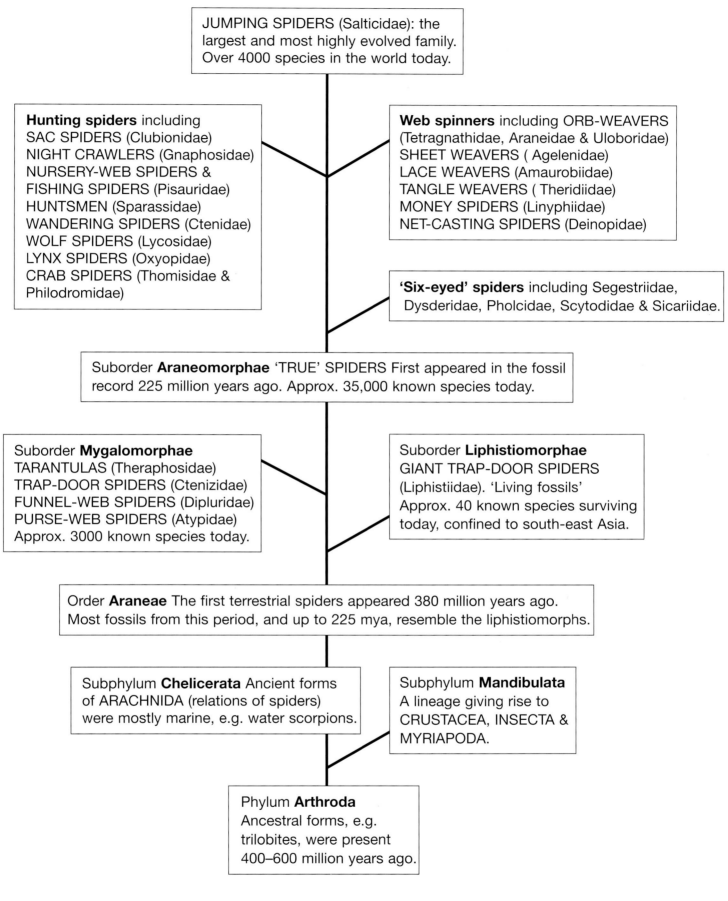

JUMPING SPIDERS (Salticidae): the largest and most highly evolved family. Over 4000 species in the world today.

Hunting spiders including
SAC SPIDERS (Clubionidae)
NIGHT CRAWLERS (Gnaphosidae)
NURSERY-WEB SPIDERS &
FISHING SPIDERS (Pisauridae)
HUNTSMEN (Sparassidae)
WANDERING SPIDERS (Ctenidae)
WOLF SPIDERS (Lycosidae)
LYNX SPIDERS (Oxyopidae)
CRAB SPIDERS (Thomisidae &
Philodromidae)

Web spinners including ORB-WEAVERS
(Tetragnathidae, Araneidae & Uloboridae)
SHEET WEAVERS (Agelenidae)
LACE WEAVERS (Amaurobiidae)
TANGLE WEAVERS (Theridiidae)
MONEY SPIDERS (Linyphiidae)
NET-CASTING SPIDERS (Deinopidae)

'Six-eyed' spiders including Segestriidae, Dysderidae, Pholcidae, Scytodidae & Sicariidae.

Suborder **Araneomorphae** 'TRUE' SPIDERS First appeared in the fossil record 225 million years ago. Approx. 35,000 known species today.

Suborder **Mygalomorphae**
TARANTULAS (Theraphosidae)
TRAP-DOOR SPIDERS (Ctenizidae)
FUNNEL-WEB SPIDERS (Dipluridae)
PURSE-WEB SPIDERS (Atypidae)
Approx. 3000 known species today.

Suborder **Liphistiomorphae**
GIANT TRAP-DOOR SPIDERS
(Liphistiidae). 'Living fossils'
Approx. 40 known species surviving today, confined to south-east Asia.

Order **Araneae** The first terrestrial spiders appeared 380 million years ago. Most fossils from this period, and up to 225 mya, resemble the liphistiomorphs.

Subphylum **Chelicerata** Ancient forms of ARACHNIDA (relations of spiders) were mostly marine, e.g. water scorpions.

Subphylum **Mandibulata**
A lineage giving rise to CRUSTACEA, INSECTA & MYRIAPODA.

Phylum **Arthroda**
Ancestral forms, e.g. trilobites, were present 400–600 million years ago.

A FAMILY TREE of SPIDERS and their ORIGINS

NOTE: This is a simplified diagram; many families are not shown. The groupings 'Hunting spiders', 'Six-eyed spiders' and 'Web spinners' do not have formal status. The chronology is from bottom to top.

HOW SPIDERS ARE NAMED AND CLASSIFIED

As with most forms of life, spiders are named on the basis of their detailed anatomy (morphology). However, in recent years, genetic profiles (DNA analyses) have become increasingly important in understanding the relationships between animals.

Under the binomial system, the name of each animal (and plant) has two parts – the genus and the species names. For example, the little black and white Zebra spider, a species of jumping spider, has the name Salticus scenicus. The first part of the name, Salticus, is the genus to which it belongs and the second part, scenicus, is its specific name.

Groups of species which are noticeably alike are given the same generic name. Among spiders, related genera, each containing one or more species, are placed in one of the 110 known families. For example, the 4,000 or so named species of jumping spiders, placed among 560 genera, are grouped in the family Salticidae (the scientific names of families always end with the suffix –idae).

Where the name of a spider consists of only a generic name, followed by the word 'species', or 'sp.' for short, it means that a spider belongs to a particular genus but it cannot be given a specific name because it is unidentifiable, or new to science.

In the overall classification of animals, spiders (both living and extinct) are represented by the order Araneae. Araneae is one of a dozen related orders within the class Arachnida. The class Arachnida, and the class Insecta, are two of the classes in the Phylum Arthropoda. The Arthropoda is one of the phyla in the Superphylum Invertebrata.

HOW TO OBSERVE SPIDERS

A good variety of spiders can be seen if you have some knowledge of their habits, but bear in mind they differ widely. For example, some are active in sunshine while others emerge at night. Spiders occur virtually everywhere, but for many people it is probably true that even the common ones remain elusive. One of the main reasons that spiders escape attention is that they have developed strategies to hide and camouflage themselves against their natural enemies. Nevertheless, it is their ability to become almost invisible that makes finding them such an exciting challenge.

Some spiders are almost the companions of humans, for many can be seen in and around the home. A wider range of species, however, may be found in a natural habitat such as undisturbed grass with occasional trees and bushes. In such a place, particularly on mornings with heavy dew, there should be an abundance of silk and webs, which is always a good sign! And if the spiders are not actually sitting in their webs, then they are probably in hiding places among the leaves nearby.

Specimens can be discovered by looking closely at flowers, grass heads, the bark of trees and frequently, the underside of leaves. Greater numbers, however, may be found by methods such as shaking the foliage of bushes over an opened umbrella, or by sweeping through grasses with a butterfly or sweep net. The best spots are likely to be where bushes are in bloom and busy with insects. For spiders living among forest litter, the best method is to scoop up a heap of dead leaves and spread them onto a white sheet. And in many places the nocturnal species, which are normally well hidden, can be seen at night in the light of a lamp, preferably one carried on the head to perceive the reflections from their tiny eyes.

While methods such as those mentioned can detect the typical species, there are others that require special searching. For example, tarantulas and trapdoor spiders are frequently hidden in burrows, and purse-web spiders need a sharp eye to spot their finger-like tubes of silk at the base of a tree trunk. Many species are specialists of particular microhabitats such as the nests of ants – unsurprisingly such spiders do require some finding! Indeed, spiders occur in all sorts of places. For example, some live unseen in the webs of larger species, others live underwater and a number lurk beneath dried cowpats.

The easiest way to catch a spider is to encourage it to move into a tube which is then gently closed with a stopper. For web-spinners, it helps to place the tube below them, as they tend to drop to the ground when disturbed. Because tubes made of glass often break during field trips, it is better to use plastic tubes. There is no need to make ventilation holes on the stoppers as spiders can easily survive for a few days in a closed tube. But they cannot stand the heat if placed in the sun and they will need a tiny drop of water. To avoid cannibalism, use a separate tube for each live spider.

Live specimens can be viewed under a lens in a glass tube and then later released if wished. If it is intended to make a permanent collection of preserved spiders, then specimens should be placed in 70-80% ethyl or isopropyl alcohol. Preserved specimens are best for permitting accurate identifications under a microscope, together with a book on identification. When making a list of species from a particular place (an inventory), it is important to preserve the specimens. It is also essential to note details of identification (if any), date, collector and locality on the labels. Labels may be printed or written with pencil or Indian ink. Place labels in the alcohol, together with the specimens, inside the tubes.

FURTHER READING

Foelix, R.F. (1996), Biology of Spiders. 2nd Edition, Oxford University Press, New York.

Hillyard, P.D. (1997), Collins Gem Spiders. HarperCollins Publishers, London.

Hubert, M. (1979), Les Araignées: Araignées de France et des pays limitrophes. Boubée, Paris.

Kaston, B.J. (1978), How to know the Spiders (Pictured Key Nature Series). McGraw-Hill Science, USA.

Levi, H.W. (2001), Spiders and their Kin. Golden Guides from St Martin's Press, USA.

Murphy, F. & J. (2000), An Introduction to the Spiders of South East Asia. Malaysian Nature Society, Kuala Lumpur.

Preston-Mafham, R. & K. (1996), The Natural History of Spiders. The Crowood Press, UK.

Roberts, M.J. (1995), Collins Field Guide to Spiders of Britain and Northern Europe. HarperCollins, London.

Simon-Brunet, B. (1994), The Silken Web A Natural History of Australian Spiders. Reed Books, Australia.

Smith, A.M. (1994), Theraphosid Spiders of the World Volume 2. Tarantulas of the USA and Mexico. Fitzgerald Publishing, UK.

USEFUL WEBSITES

American Arachnological Society www.americanarachnology.org/

British Arachnological Society www.britishspiders.org.uk

British Tarantula Society www.thebts.co.uk

Groupe d'études des Arachnides http://gea.free.fr/

The Arachnology Homepage www.arachnology.be/Arachnology.html

The World of Jumping Spiders www.salticidae.de/index1.htm

The World Spider Catalog, Ver. 7.0 www.research.amnh.org/entomology/spiders/catalog

Spider Bites in the UK www.nhm.ac.uk/nature-online/life/insects-spiders/spiderbites/

INDEX

PICTURE CREDITS

A.N.T. Photo Library/NHPA: 26/7, 58; Ingi Agnarsson: 49, 124/5, 126, 127; Ingo Arndt/nature-pl.com: 10, 13, 24, 70(l), 70(r), 71(l), 71(r); Steve Aylward: 147; Anthony Bannister/NHPA: 23, 29, 108, 129; Alan Barnes/NHPA: 57, 105; Niall Benvie/naturepl.com: 76/7; George Bernard/NHPA: 25, 103; N. Callow/NHPA: 2; John Cancalosi/naturepl.com: 22; James Carmichael/NHPA: cover (back flap), 1, 5/6, 7(b), 15, 17, 20, 28, 33, 34, 38, 65, 78, 85, 98/99 100, 101, 110/1, 112, 114, 136; Stephen Dalton/NHPA: cover (front flap), 7(t), 7(m), 8/9, 11, 31, 36/7, 39, 40, 41, 42, 43, 44/5, 46, 47, 48, 50, 52, 55, 59, 62/3, 72, 79, 82, 89, 97, 106/7, 121, 122, 138, 148/9; Manfred Danegger/NHPA: 139; Bruce Davidson/naturepl.com: 120; Nick Garbutt/naturepl.com: 67; Nick Garbutt/NHPA: cover (back), 6, 64; Sébastien Genevieve: 116; Ken Griffiths/NHPA: 117; Adrian Hepworth/NHPA: 21; Daniel Heuclin/NHPA: 16, 66, 68, 69, 75, 88, 93, 94, 95, 104, 115, 118, 119; Ernie Janes/NHPA: 144; Hans Christoph Kappel.naturepl.com: 73; George McCarthy/naturepl.com: 84; Meul/ARCO/naturepl.com: 90/1; Claus Meyer/FLPA: 131; Mark Moffett/FLPA: 133; Ross Nolly/NHPA: cover (front), 30; Rolf Nussbaumer/naturepl.com: 53; Rod Planck/NHPA: 32, 109; Premaphotos/naturepl.com: 35, 86/7, 92, 96; Robert Thompson/NHPA: 51; John Shaw/NHPA: 87; Adrian Shepherd: 74; Lynn M. Stone/naturepl.com: 80/1; Kim Taylor/naturepl.com: 12; Roger Tidman/NHPA: 145; Anne & Steve Toon/NHPA: 3, 83; Wegner/ARCO/naturepl.com: 134/5.

ACKNOWLEDGEMENTS

Thanks are due to Andrew Smith, the President of the British Tarantula Society, for his generosity in providing nuggets of information. Miranda Hillyard and William Hillyard (niece and nephew) are both thanked for their constructive comments on the draft and Robin Hillyard (brother) for his help with the glossary. Appreciation is also due to Oliver Crundall for his inspiration as a spider detective and his ability to spot specimens that I had missed. Finally, thanks go to my wife, Leni, for her vital support, and daughter Mia, for not disturbing me too much!